PLENTY AND TROUBLE

The Impact of Technology on People

Robert H. Hamill

Nashville • ABINGDON PRESS • New York

PLENTY AND TROUBLE

Copyright © 1971 by Abingdon Press

All rights in this book are reserved.
No part of the book may be reproduced in any manner whatsoever without written permission of the publishers except brief quotations embodied in critical articles or reviews. For information address Abingdon Press, Nashville, Tennessee

ISBN 0-687-31655-3

Library of Congress Catalog Card Number: 73-148066

Scripture quotations unless otherwise noted are from the Revised Standard Version of the Bible, copyrighted 1946 and 1952 by the Division of Christian Education, National Council of Churches, and are used by permission.

SET UP, PRINTED, AND BOUND BY THE
PARTHENON PRESS, AT NASHVILLE,
TENNESSEE, UNITED STATES OF AMERICA

*To the one named Deda
and the four named Den, Tim, Gwen and Greg
who share with me all this plenty
and most of this trouble*

Contents

Preface 7

1 Surrounded by the Past and the Future ... 11

2 The Technological Spirit Today 28

3 Creatures in the World of Nature 42

4 What Is Man More Than a Machine? 66

5 Self-made Men 93

6 Some Pro's and Con's 132

7 A Christian Critique 153

Suggested Reading 185

Index 190

Preface

It all began with the wheel, or the first little cooking fire, or the time a man used a stick to pry loose a boulder. The first tool—that's when technology began, according to simple definition. "The totality of tools" is technology, says Emmanuel G. Mesthene, and the American toolbox includes a wide assortment of hand tools and machines and the spectrum of intellectual tools, including computer language, the scientific method, and ideas about research, planning, and budgeting.

What can the sensitive Christian person think and feel about this technological time? There are three facile, one-dimensional views we must discard. Some people jump up and down with excitement because they see technology as pure blessing, the dynamo of all progress. A second view sees technology as unmitigated curse. Isaac Asimov does not expect mankind to make it to the year

PLENTY AND TROUBLE

2000. "This is the first generation which has a choice of suicides. It can squeeze itself to death, poison itself, or blow itself to bits." By this view, technology is a monster let loose, and technological man a planetary disease. People who have been hurt or frightened by the technological changes succumb to this view with a kind of perverse rejoicing: "We are doomed, hallelujah!" The third view argues that there is nothing special about the technological revolution; it really is not a revolution, just an intensified movement in the direction we have been moving for several centuries. And just as we survived the factory, so we will survive the computer. Don't fret. There's nothing here to worry about.

Three oversimplified attitudes: the dance, the scream, and the yawn. Unhappily, none of these recognizes the penetrating impact technology makes on the way we think about all things, including technology itself. Technology covers an enormous range of experiences, far far surpassing anything Johann Beckmann could imagine when he coined the word in 1772. This book explores only a few examples of the impact technology makes on people. It makes hardly any mention of mass media, for instance, or education, or industrial management, highways, advertising, agriculture, nuclear power, weapons, fallout, the invasion of privacy, or the family and sex. It does

PREFACE

plunge into a few large areas where technology makes severe impact on daily life: ecology, the computer, bio-medical experiences, and the pervasive experience felt as tension or pressure. I intend to explore these matters from the standpoint of a Christian critique. I am concerned with the Christian understanding of creation and the natural world. I take seriously both the grandeur and the misery of human nature: your capacity to achieve good things and my capacity to pervert any good thing you can do. I find in the Bible the record of God's struggle to overcome the human hankering to go backward, to stand pat, to settle down. I come out convinced that technology is here to stay and will increasingly penetrate our whole way of living and thinking. Technology, like the original Garden, contains apples and snakes. The apple is too tempting to resist, the snake too devilish to escape. This book simply invites you to reflect seriously on what you are doing and thinking and feeling today, and what kind of person you will be tomorrow, if you don't watch out.

Gladly I express my thanks to Boston University for granting me a leave of absence to pursue this study; to a good number of people in England and Scotland, including Miss Mollie Batten, Jeremy W. Bray, M.P., Brian L. Cordingley, Gordon Dunstan, William Gowland, Kenneth G.

PLENTY AND TROUBLE

Greet, George D. Wilkie, and especially Dr. David Edge, Chairman of the Science Studies Unit at the University of Edinburgh, all of whom talked with me about these matters out of their experience and study; and to Dr. Emmanuel G. Mesthene and Scott Paradise, of Cambridge, for their several conversations; and to colleagues here at Boston University. This book has developed out of lectures given originally on the Voigt Lectureship on Preaching at McKendree College; I am especially thankful for that invitation, and for the hospitality and the critical hearing that I received.

Mrs. Beth Ballard and Mrs. Mary Simpson typed the manuscript; Mrs. Dorothea Overholt corrected the galley proofs; and my wife kept reminding me of the Protestant work ethic, meanwhile using her library experience to stack upon my desk many books I have yet to read, alas!

To all these people I say thanks, and absolve them of any responsibility for errors of thought or fact that may yet be found here.

R. H. H.

Marsh Chapel
Boston University
Boston, Massachusetts

1

Surrounded by the Past and the Future

Men distort their pasts to make them conform to their present excitements and serve their present interests. In our day we are fascinated by machines, and we consider it the purpose of life to use machines to control nature and develop the environment. Therefore we suppose that our ancestors were equally consumed by this effort to make and use tools. Hence we twist our history to make it look like a straight line from our ancestors to ourselves as technological men.[1]

Indeed, human history can be written in terms of human tools. It has been done,[2] and it makes a fascinating story. As early men evolved they made cutting tools from flint and the hand ax from

[1] See Lewis Mumford, *The Myth of the Machine* (New York: Harcourt, Brace & World, 1966), p. 14.

[2] T. K. Derry and Trevor Williams, *A Short History of Technology* (Oxford: Clarendon Press, 1961), a condensation of their five-volume work, *History of Technology*.

PLENTY AND TROUBLE

quartz or lava. To hunt for dinner they made the spear and the fishhook, then the club and the bow to match their skills against the brute strength of the bear and the ox. When they settled down in little communes they carved out a hoe, then a plow drawn by the ox; then they put wheels under a box and had a wagon for hauling. In due time they put a sail onto a hollowed-out log and had a sailing vessel. They took to the seas and began to trade their pottery for cedars of Lebanon. For weapons they "improved" the bow and arrow by developing a catapult and heavy ballista; then the battering ram made war profitable, and the military-industrial complex arose! The Romans, the first major technologists, invented interior plumbing, then they built aqueducts that still stand in southern France, and roads, and factories for making cloth and armament. Meanwhile, sculpture tools made possible "the beauty that was Greece, the grandeur that was Rome." During the Middle Ages in the Far East, steel, silk, paper, porcelain, Arabic numerals, scientific inquiry, chemistry, and universities—such were the tools of men's making. Know the history of tools and you know a great deal about human development.

At the beginning of this study we do well to warn ourselves against reading human history from the physical artifacts only. That is the folly of archaeology. It overemphasizes tool-using because

SURROUNDED BY THE PAST AND THE FUTURE

it must confine its evidence to material goods. Only physical tools can endure centuries of burial under the hot sands. Language, ritual and ceremony, social structure, family life, sport and play —these leave no visible evidence. Only knives and coins and crockery. Thomas Carlyle described man as a tool-using animal. Mumford writes:

This overweighting of tools, weapons, physical apparatus, and machines has obscured the actual path of human development. . . . There was nothing uniquely human in toolmaking until it was modified by linguistic symbols, esthetic designs, and socially transmitted knowledge. At that point the human brain, not just the hand, was what made a profound difference.[3]

Man is *homo faber*, to be sure, but not only *homo faber*, and not primarily *homo faber*. This book, however, deals with technology, therefore with tools and machines, although technology is more than machines. Therefore we look now at some of the "hardware" of technology, the visible tools men make and use. I choose five such tools, five crucial innovations which illustrate the enormous impact that technological changes make on the daily lives of people.

1. *Fire.* It may seem strange to list fire among

[3] Mumford, *The Myth of the Machine*, p. 5.

PLENTY AND TROUBLE

the visible, durable "tools" of technology, but give attention first to Mumford's eulogy of man's technological use of fire.

Language apart, [fire] counts as man's unique technological achievement: unparalleled in any other species. Other creatures use tools, construct dwellings, dams, bridges and tunnels, [they] swim, fly, perform rituals, cooperate as families in raising the young; and even, among the social ants, wage war with soldiers, domesticate other species, and plant gardens. But man alone dared play with fire: so he learned to court danger and to discipline his own fears; and both practices must have enormously increased his self-confidence and effective mastery.[4]

This tribute illustrates how tools make their immense impact on people. Fire gave men "effective mastery" over their environment; it made the winters livable and the beefsteak edible, and it gave men a good night's sleep by keeping the wolves away. But more than that, fire enticed men to play with danger. It gave them the exultant sense of conquering their fears of a new demon. Is it too much to say that fire made men out of beasts? Fire had that profound impact. To ignite it, to spread it, to tame it, to make it work for you—that was the first vivid technological experience of human beings.

[4] *Ibid.*, p. 124.

SURROUNDED BY THE PAST AND THE FUTURE

2. *Glass.* In a cursory glance at human inventions one might not consider glass especially notable, except for two profound consequences. When glass was refined into eyeglasses, it prolonged the intellectual life an average of fifteen years for the normal person. That just about doubled the productive years of the person who read and wrote. The Renaissance revival of learning resulted partially from this spectacular increase in human productivity. Then glass was made also into the magnifying glass and the compound lens. This created the whole science of optics by which the microscope and telescope drastically changed the dimensions of the known world. The cosmos opened up its incredible size and majesty, and the tiny world of organic change and chemical mysteries became visible. Superficial observations gave way to exact study, and science became possible.

3. *Printing.* Gutenberg's movable type and printing press provide the classic illustration of the social impact of technological innovation. Printing broke down the elitist concept of education by making the printed page available to everyone, including the poor. The printing press democratized knowledge and thus contributed to the social turmoil that followed the breakup of the medieval class system. As soon as people could read "We

PLENTY AND TROUBLE

must obey God rather than men," they began to do exactly that, to the discomfort of kings and governors. Printing also illustrates the reinforcing effect of technological change. When reports of scientific discoveries were printed and widely circulated, that increased the scope and the tempo of more discovery, hence, more and faster mechanical invention. Thus the technological revolution worked up a head of steam.

4. *The clock.* Sundials, the hourglass, the water clock, and other devices testify to the human fascination with time. From early times men wanted to measure time, but they could make only a sloppy approximation until the mechanical clock was developed in the fourteenth century. Of all human inventions, the clock has since become the most technological in spirit. The clock regulates the day's activity and sets bounds to the common life. The clock synchronizes human relations so that everyone arrives at the committee meeting "on time"—hopefully. The clock introduced exact measurement and temporal control into almost every activity. Minimum wage is so much *per hour,* vacations are two *weeks,* and retirement is set at *age* 65. The clock tyrannizes. (I am writing this chapter while on vacation, and even here my wife—Eve took the apple!—makes the regime for the day by the clock. I am to start

work at 7:30, and no sunning until 11:00 A.M.!) The clock is the model for all instruments of precision. How could NASA foretell the moment and location of a moon-landing without exact measurement of time, down to the fraction of a second? The model machine in our society is not the gigantic Saturn rocket but the miniaturized clock.

The clock illustrates beautifully the profound effect of technology upon our thinking and understanding of nontechnological things.

It is in the works of the great ecclesiastic and mathematician Nicholas Oresmus, who died in 1382 as Bishop of Lisieux, that we first find the metaphor of the universe as a vast mechanical clock created and set running by God so that "all the wheels move as harmoniously as possible." It was a notion with a future: eventually the metaphor became a metaphysics.[5]

5. *Railroad.* When government planners tried to anticipate the impact the space program might have on American society they looked about for some historical parallel, and settled on the railroad for a careful study. Their research clarified the enormous repercussions resulting from the "iron horse." [6]

[5] Lynn White, Jr., *Medieval Technology and Social Change* (London: Oxford University Press, 1962), p. 125.

[6] Bruce Mazlish, ed., *The Railroad and the Space Program* (Cambridge: The M.I.T. Press, 1965).

PLENTY AND TROUBLE

(1) Clearly the railroad gave the country a new transportation system. This meant a more constant supply of goods, with therefore less fluctuation in price, which benefited the farmers. Refrigerator cars meant fresher food, a benefit to the consumer. (2) The railroad meant travel. People took vacations and visited their relatives and the national parks. Hotels and resorts prospered, fairs and expositions and baseball games all benefited. Salesmen took to the parlor cars to cover their territories, and who can estimate the influence of the traveling salesman upon the instability of the American family? (3) Railroad tracks bridged over the valleys and tunneled through the hills. Cars kept goods warm in winter and cool in summer. The ravages of nature were checked. (4) Adventuring pioneers expanded the frontier and brought marginal land into profitable use. (5) In turn the railroad brought people into the city and made urban life more dense and labor more concentrated, resulting in the "city problem" and the "suburban problem" and the whole concept of the "right and wrong side of the tracks."

(6) The railroad stimulated other technical changes. It created the need for the telegraph, and air brakes, and the new Bessemer steel, and it consumed a gigantic amount of coal, resulting in vast changes in the coal and iron and communications industries. (7) The huge size of the

SURROUNDED BY THE PAST AND THE FUTURE

railroad enterprise required a new administration. Financiers, engineers, and management bureaucrats rose to control in place of owners. That contributed to the gulf between management and labor, resulting in the origin of labor unions. In turn that brought government mediation, subsidies, rate-setting, and franchises.

(8) This cursory listing of the impact of the railroad must include its effect upon the pastoral image, the image of innocent rural life being consumed by the aggressive iron horse. The masculine, dynamic locomotive devoured the idyllic vision of purple mountains and amber waves of grain. In the conflict of city against country, the railroad made the industrial process of the city so dominant over the agricultural process of the farmland that America became a city culture and nature was almost obliterated. The railroad remade America. That is the mildest assessment possible.

Fire and glass and printing and the clock and the railroad—these human tools, the "hardware" kind of technology—account partly for the shape and the feel of our society. They belong to our history, but they are very much with us today and tomorrow.

Therefore let us take a serious look into the future, which in some ways is already here.

Science fiction writers whet our appetites for the future. They describe a world once thought to

PLENTY AND TROUBLE

be incredible, yet each passing year brings closer those wild imaginings about the future. We moderns seem intoxicated by a fascination for the future, and we tend to see it full of hope, a promised land of health and wealth and all things good. "Just as the approach of the year 1000 AD. created a kind of millennial madness in Christendom, so the approach of the year 2000 seems to be producing a kind of secular millennialism," says Kenneth Boulding.

In fact, the name of the game is Future. Almost everyone plays the game, today as in times past. Future-guessing occupies the human imagination. Why else did the Greeks go to the oracles? Why did men turn to astrologers? Why the horoscopes in the newspapers, and the songs and dress patterns and art using the signs of the Zodiac, except for our anxiety about the future? The crystal ball holds us entranced, and for good reason, because the image we hold of the future helps to shape the future. What we expect to happen is more likely to happen—more likely than if it were not expected. Expectation takes on the feeling of inevitability. Hence we bend toward what we feel is "in the cards."

Future-guess is not a game only. Thoughtful men make predictions of things likely to happen. That ominous year 1984 looms just ahead, and sober men wonder about Huxley's and Orwell's

SURROUNDED BY THE PAST AND THE FUTURE

nightmarish visions. The American Academy of Arts and Sciences set up a Commission on the Year 2000, and Herman Kahn and his associates in the "war game" exercises have published a book, *The Year 2000: A Framework for Speculation on the Next Thirty-three Years.* The Foreign Policy Association upped the ante with a book, *Toward the Year 2018*, and there is (was!) a popular song about 2020! Books, study groups, federal grants, projections by planners in business and government and universities—such are the works of a new breed of "futurologists." Their publications include a magazine, *The Futurist,* called a newsletter for tomorrow's world. Indeed, "prognostics" has become a major indoor sport.

It makes sense to anticipate. If population is to double, someone had better build more schools and stir up the housing industry. If computers are to replace clerks and accountants, then clerks and accountants had better hurry to learn new skills. If . . . then. . . . Sensible people are rightly anxious about the future.

Prediction of the future depends on what Boulding calls the wallpaper principle. We perceive the stable and repeating patterns in current events, then foretell the future in somewhat the same way as we calculate the pattern of the wallpaper behind the mirror that obscures it. We project the trends and velocities of social change and get an

PLENTY AND TROUBLE

approximation of the future. Elementary, Watson. Yet that is the substance of "prognostics." Let's play the game ourselves.

Travel. Adam walked the good earth in Eden; Neil Armstrong took a giant step for mankind on the cold surface of the moon. Therefore the space people predict there will be manned bases on the moon by 1980, and in due time commercial passenger rockets to the moon. Manned spacecraft will land on Mars and Venus by 1990, and by the year 2000 men will live on artificial satellites for extended periods of time for scientific research.

By the year 2000 earth travel will include electronic control of the private automobile, itself battery-powered, permitting the driver to dial his destination and then let "the system" take over his driving, speed, braking, and all. In the air, supersonic travel will be customary, and some form of vertical takeoff and landing craft will capture the short-distance traffic.

Computers. Already we have computers that perform almost incredible tasks. They can count, keep records, translate, do research and other tasks with unbelievable speed and accuracy. They make up corporation payrolls and keep inventories; they record the nation's tax returns, diagnose disease, operate factories, control air traffic, read and write, learn and teach, play chess and even play

SURROUNDED BY THE PAST AND THE FUTURE

Cupid, as students know who use the mate-matching schemes. The day is coming when a household will have a console where the housewife will check prices and place orders at the neighboring stores, do her banking, order specific television shows, and for diversion attain an advanced degree at the university—all via computer at home.

Already the computer charges the coffee an employee takes from a vending machine and deducts it from his paycheck. Someday it will accept a person's church contribution, so instead of dipping his fingers into the holy water, he will insert his credit card into the machine in the narthex and have his contribution deducted from his bank account. That will do away with the offering and instead give an instant count of the day's contributions, which could then be announced or flashed on the screen under the pulpit. For televised services, a message could flash across the home screen telling people to dial their bank and have their funds transferred to the church account.

One of the major political parties has set up an automated system to keep track of voters. It has stored over six million names, with each person's age, sex, religion, occupation, family status, and level of income duly noted. The computer records also whether the person works for the party, or

PLENTY AND TROUBLE

is a veteran, or a union member, or a corporation executive, or belongs to the PTA. Then it can type out a "personalized" letter, at the rate of 48 lines in six seconds, and run off address tapes at the rate of 600 per minute.

A related development will be the phonovision and portable television, which exist already; in combination these will make it possible for several people to hold a "conference" while miles apart. That will reduce the need to travel to the office for work, and thus ease the traffic and pollution problems and save time from travel.

Education. A national data center, connecting libraries and laboratories, will give out any desired information through monitoring devices, so that on the private console at home a person can get stock exchange reports and today's sports and comics, as well as historical studies and scholarly research. Education will include knowledge accumulated in electronic banks, and in the distant future, machines may transmit information directly to the human brain through coded, electronic messages, which means you don't have to study or listen to sermons or lectures. You simply plug your brain into the computer and it gives you instantaneously whatever you want to know!

Computers will analyze handwriting and transcribe it, and analyze the human voice and tran-

SURROUNDED BY THE PAST AND THE FUTURE

scribe it onto paper; other computers will translate one language into another. By hooking these machines together a person might scribble a note, put it under a photo-electric cell, and have it typed out in Athens in Greek. Or an American could dial Tokyo and speak English to a Japanese friend who hears it in Japanese; he then answers in Japanese and the American hears it in English. (Do we really want to understand each other that well?)

Already there is a machine to help the blind to read. It uses a small hand-cell to scan the printed page. It relays impulses through a grid of 144 tiny circuits to a plate where the reader places his finger and receives electric impulses something as though he were reading Braille, except that this machine reads any printed page!

Food and energy. New foods will be developed by "farming" the ocean and by synthetic processing in the laboratory. Energy will be captured from the sea and the sun to reduce human toil. Minimum control of the weather will eliminate drought and flood, and sea water will be desalinized for drinking and irrigation.

Bio-medicine. In the field of genetics Ellul predicts there will be a stable population of the highest types, achieved by effective fertility control and by conception outside the human body

PLENTY AND TROUBLE

with sperm and ova taken from persons of superior heredity. Organ transplantation will be supplemented by the development of artificial organs; after all, the heart is just a small pump! Many hereditary defects will be erased by genetic engineering. Non-narcotic drugs will be used to change personality, to increase intelligence, and to increase life span by thirty years. Primitive forms of life will be created artificially, and conceivably something resembling human life will be nurtured in the laboratory.

A Christian Heresy

Such predictions of things to come express a secular hope. The year 2000 will celebrate the new age! Spies in the Promised Land bring back extravagant reports of a land flowing with milk and honey. The future skies look blue, yet many people see ominous clouds overhanging. We postpone any judgment. In this chapter I simply look into the crystal ball to see what is possible and likely. Whether it is humane or demonic, whether the Christian will see it with cheer or with dread—let us delay those questions until later.

There is, however, a Christian coloration to all this. According to the ancient religions, the world moved in cycles. Just as the seasons turned each year, so human life was thought to be bound to

SURROUNDED BY THE PAST AND THE FUTURE

the unbreakable cycle of birth-life-death. Christianity shattered that view and put human affairs on a straight line that began with Creation, moved on through Redemption and headed toward the Last Judgment, with some thought given to the Final Coming. This conception of things got secularized into a belief in progress and the accumulative achievement of the Kingdom of God on earth, but the point is that this whole conception of a good future, the dream of realizing human hopes, arose out of the Christian conviction that the future has a future, it is not a repetition of the past. What John Oldham once said about the hope in progress can be as truly said about technology and its promise: it is "a Christian thing in the sense that only in what has been Christendom could it establish so firm a grip upon the hearts of men. In other words, it is a Christian heresy and not a false doctrine." As we move along in this study we will probe further into this idea that technology, because of its grandiose vision of the future, is a Christian heresy. For the moment we have simply taken snapshots of the future. Let us turn now to the present and probe a bit into the fundamental nature of technology as we experience it today.

2
The Technological Spirit Today

The authorities are not agreed (par for the course!) about the origin of the word "technology."[1] In any case it is a recent word that connotes the complex condition that characterizes the last two centuries of the western world. However we come to define it, technology gives the shape and the distinctive flavor to modern American life. We are a technological people. But what is that? And what's new about it? Have we not been heading this way for at least two hundred years? What is the present but an intensified, up-dated version of past trends? More speed, more power, more people, more problems, no doubt—but

[1] Sir Robert Watson-Watts credits Johann Beckmann with the first use of the word in 1772. John G. Burke declared, "The word *technology* was the creation of Prof. Jacob Bieglow of Harvard, whose *Elements of Technology* in 1829 brought the word into the English language." Prof. Lynn White gives the still later date of 1873.

THE TECHNOLOGICAL SPIRIT TODAY

"more" of anything is not equal to anything "new."

We need beware of that slippery logic, however. "More" cases of typhoid do make a "new" epidemic. "More" officeholders constitute a "new" bureaucracy. "More" people turn the countryside into a city, and a city is distinctly different from a farm. Enough quantity becomes different quality. Therefore to the question, Is there really anything new under the sun? I answer Yes. The shadings of difference in modern life add up to a new thing. The technological nature of our life makes a new thing under the sun.

Novelty in Modern Technology

To get acquainted with technology we devoted chapter 1 to our technological ancestors, and we speculated about our descendants. Now we look at ourselves: how we feel and act and think as technological people. We do this by checking off the novelties in the technology of our times.

1. *The skyrocketing graph.* When a graph of such matters as speed or population or energy consumed is plotted on a base line running left to right across a long sheet of paper representing the time span of all human history, the graphed line hugs the base line closely until suddenly, mid-nineteenth century, it rises sharply, and then in our lifetime it spurts dramatically upward, almost

PLENTY AND TROUBLE

vertically, and shoots far off the page and out of sight.

Speed, for instance. For many millions of years a man's top speed was his own walk or run. On his own he still is limited to fifteen miles per hour, after all these centuries! Then on camel he made eight miles an hour, and on horseback twenty miles per hour in a chariot race. The mail coach of colonial days and the great sailing ships made less than ten miles per hour, and the first steam locomotive could accomplish only thirteen. Late in the nineteenth century the roaring iron horse made one hundred miles per hour. It had taken the human race millions of years to attain that speed. But then within the next fifty years the airborne man multiplied that four times, and by 1960 his rocket planes multiplied that by ten and reached the speed of 4000 mph. Then beyond gravity the space men circled the earth and headed for the moon at 18,000 mph. Within one generation the graph line shot up far off the page.

Distance, also. A man's voice was limited to how far he could holler. The echo might be heard five miles down the rock gulley. Then only one century ago Alexander Graham Bell transmitted the voice across the country. Fifty years ago Marconi liberated it from wires and let it loose wherever a crystal could pick it up. Then in July, 1969, we heard Neil Armstrong say, "One step for man. A

THE TECHNOLOGICAL SPIRIT TODAY

giant step for mankind," across 225,000 miles. The graph line skyrocketed out of sight.

The same is true about calculation. For millions of years men counted on fingers and toes, and with sticks and stones. Far toward the right-hand edge of the graph the Chinese invented the abacus, and someone conceived of logarithms, and someone else invented the calculator, but it was still a pencil and paper arithmetic until suddenly, in 1945, came the computer, and when its transistors replaced vacuum tubes the machine could handle fourteen million operations per second!

Thus it was with many other measures of physical command. Centuries went by with imperceptible change. Then in our lifetime came a sudden acceleration of change that sent the graph spurting upward, almost beyond comprehension. That skyrocketing rate of change is one thing new about technology and the way we live.

2. Another is *the spectacular growth of knowledge*. Man's accumulation of useful knowledge has been spiraling upward for thousands of years. It took a leap upward with the invention of writing but remained painfully slow until the printing press. Before 1500, estimates say, Europe was producing not over a thousand books a year, but in our time some one thousand books are published each day. (See Eccles. 12:12) Scientific journals have multiplied astronomically. Some 1000 pro-

PLENTY AND TROUBLE

fessional journals were published, on all matters, in 1850. By the turn of the century this figure had multiplied by ten, to 10,000. By 1950 it became 100,000, and by the mid-sixties there were over one million professional journals published regularly around the earth. No wonder the scientists say they cannot keep up with the work in their own fields; they cannot read even the digests; they need a disgest of the digests! Such a knowledge explosion has never happened before, and especially the deliberate pursuit of knowledge as a quest for power. Useful knowledge is no longer a delight-for-its-own-sake, nor the pride and privilege of the educated man. Now knowledge is considered the fuel of progress. People grasp for it as the miser grabs for gold. Knowledge equals power. This is something new under the sun.

3. *Technology is based on science laboratories.* This novelty makes another distinctive feature in our times. It means, simply put, that we have created a new way of doing new things. Whitehead said it vividly, "The greatest invention of the nineteenth century was the invention of the method of invention."

Innovations in technology depend not upon some obscure or pervasive force but upon the actions and decisions of individual men, and increasingly it is the scientists, the scientists-turned-practitioners, who do the innovating. The growing merger of science and technology results from the Research and Development divisions in industry, where "science" pursues research for the purpose of development. Massive government grants and foundation gifts pour into the university science departments to commit the academic laboratories to innovation, and "pure" research often becomes adjunct to some profitable objective. "Part and parcel of the idea of *scientific* progress is that we investigate reality.... In *technology* we do not investigate reality but create a reality according to our own designs.... Scientific progress is the pursuit of knowledge; technological progress is the pursuit of efficiency."[2] Such is the ideal, the definitive distinction between science and technology, but increasingly the scientist has become a developer. He sets out deliberately to develop a product or process to serve some specified need. Thereby, according to Jacques Ellul, the profound student of this whole matter, the scientist has "deliberately oriented his research toward the scientific discovery that would be applied technically.... In the twentieth century, this relationship between scientific research and technical invention resulted in the enslavement of science to technique."[3] "Enslavement" is a harsh

[2] Henry Skolimowski in *Center Diary 18,* (Santa Barbara, Calif.: The Center for the Study of Democratic Institutions), pp. 56-57. Italics added.

[3] *The Technological Society,* trans. John Wilkinson; intro-

PLENTY AND TROUBLE

word, but Ellul believes that the close link between the science lab and the industrial technology has chained the scientist to the technological spirit. He searches for knowledge in order to develop something practical and efficient. Research becomes systematic and goal-directed. Whitehead calls it invention, Ellul judges it enslavement. In any case it is part of the new thing called modern technology.

4. Another new dimension of modern technology is *the shortened time from first idea to practical use*. The period between original concept and consumer use has been drastically reduced. Formerly it took a long time for a new discovery to get into popular use. It took centuries, for instance, to utilize Paracelsus' discovery that ether makes a good anesthetic. In 1836 a machine was developed that could mow, thresh, tie straw into bundles, and pour grain into sacks, but the combine was not marketed until a century later. The first patent for a typewriter was issued in 1714, yet none appeared on the market until after the Civil War. Such delays between idea and application are unthinkable today. We may be no more energetic than our ancestors, but we know how to get ideas marketed much faster. Whereas it took 102 years for photography to reach the public, 56

---
duction by Robert K. Merton (New York: Alfred A. Knopf, 1964), p. 45.

THE TECHNOLOGICAL SPIRIT TODAY

for the telephone, 25 for X rays, in more recent times it has taken only 10 years for nuclear reactor, 5 for radar, 3 for the transistor. Technological innovations reinforce one another and the whole process intensifies, something like a tetherball game. Once you get the ball swinging your way, its radius gets shorter and its cycles get faster.

5. This leads us to still another distinctive feature of modern technology: *innovations often have a pervasive and rapid effect on society.* Television, for instance, has enormous impact on family life, school instruction, political campaigning, business advertising, mores and morals. Modern technology is so pervasive that it has become qualitatively different from the technology of the past.

It becomes clear, therefore, that modern technology is a new thing under the sun. The characteristics of earlier technology no longer prevail. Formerly, new tools were pathetically imperfect, they developed slowly and were confined to local use, and people were free to use them or not. Now everything is different.

Today's technical phenomenon has almost nothing in common with the technical phenomenon of the past. . . . It has been extended to all spheres and encompasses every activity, including human activities. It has led to a multiplication of means without limit. It has perfected the instruments. . . . Technique has extended geographically so that it covers the whole earth. . . . We are faced with the exact opposite of the traits previously in force.[4]

Technology has two other characteristics worth mentioning at this point: it assumes that nature is meant to be exploited for the satisfaction of human desires, and it emphasizes quantity "as the key to the true and measure of the good." (Ferkiss)

Before we try a definition of technology, we owe it to the subject to observe that perhaps the most distinctive characteristic of technology is change. It sounds like a cliché to say that change is the only constant, but cliché or not, it feels true. People feel assaulted by unceasing change. They experience what Alvin Toffler calls "future shock" because a new society is erupting in their midst, a fast-paced, fragmented society filled with bizarre styles, new choices, and strange customs.

An alien culture is swiftly displacing the one in which most of us have our roots. Change is avalanching upon our heads, and most people are unprepared to cope with it. . . . In certain quarters, the rate of change is already blinding. Yet there are powerful reasons to believe that we are only at the beginning of the accelerative curve. History itself is speeding up.[5]

[4] *Ibid.*, p. 78.
[5] "Future Shock," *Playboy*, February, 1970, p. 97. Two articles later developed into a book by the same title.

THE TECHNOLOGICAL SPIRIT TODAY

So much so that one authority declares, "No exaggeration, no hyperbole, no outrage can realistically describe the extent and pace of change ... In fact, only the exaggerations appear to be true." Toffler goes on to describe the *feel* of change, the computer he found in a Paris convent for instance, and the fast-food, stand-up, gulp-and-go lunch counter, and (in my case) the chainsaw that wakened me one morning up in the Michigan woods. The electric eye opens the door, the ten-minute parking meter keeps cars on the move, and shuttle planes between major cities run as streetcars used to do. Consequently, people feel transient, restless, on the move. More than half of the 885,000 listings in the Washington D.C. telephone book for 1969 were new from the year before. Toffler mentions also the temporary organizations (the popularity of *ad hoc* task forces) and the impermanence of knowledge (half of what a science student learns is outdated within ten years.) "Things, places, people, organizations and information will speed through his life, compelling him to learn, dislearn and relearn, to commit and uncommit himself, to adapt and readapt—in short, to *live*—at a faster pace than ever before." [6]

All this speed and change constitutes a major "fact" about technology. This is what technology is all about.

[6] "Future Shock," *Playboy,* March, 1970, p. 89.

PLENTY AND TROUBLE

A Try at a Definition

Technology begins with "hardware," the physical goods ranging from the geodesic dome and Saturn rocket to the Pill, the computer, the blood analysis machine, the garbage disposal, the traffic control signals, the genetic experiment lab, TV, DDT and SST. It is not enough, however, to define technology as the use of tools, not even the systematic use of tools. W. S. Roberton comes closer. 'Technology is the systematic use of systematically acquired knowledge by systematically educated people."

A profounder understanding comes from Prof. Mesthene, who is one of America's most distinguished scholars on this subject, and head of the Harvard University Program on Technology and Society.

We have found it more useful to define technology as tools in a general sense, including machines, but also including such intellectual tools as computer languages and contemporary analytic and mathematical techniques. That is, we define technology as the organization of knowledge for the achievement of practical purposes.[7]

[7] Emmanuel G. Mesthene, *Technological Change* (Cambridge: Harvard University Press, 1970), p. 25. Peter Drucker says it briefly: "Technology is not about tools; it deals with how man works."

THE TECHNOLOGICAL SPIRIT TODAY

Mesthene here introduces into our definition the element of purpose. Technology aims to do certain specified tasks. It has "practical purposes." Also it includes the intangibles of linguistic skills and intellectual methods and certain new "techniques" of analysis and mathematics. Technology thus includes the urban affairs office, the army manual, weather forecasts, the engineers' professional code, the airport control tower instructions—things and people and concepts which the trade calls "software" as distinct from the "hardware" of machines.

The most profound analysis of technology in our time has been done by Jacques Ellul, an articulate Christian layman and professor at the University of Bordeaux. In his book, *The Technological Society*, a major work everyone must read who wants to understand technology, Ellul uses the word "technique" to cover what we call technology. In the foreword of that book, Robert K. Merton makes it clear that

by *technique* [Ellul] means far more than machine technology. Technique refers to any complex of standardized means for attaining a predetermined result. Thus it converts spontaneous and unreflective behavior into behavior that is deliberate and rationalized. The Technical Man is fascinated by results, by the immediate consequences of setting standardized devices into motion.[8]

[8] P. vi.

PLENTY AND TROUBLE

According to Lewis Mumford, this technique began with the Egyptians, when Pharaoh organized his slaves into a gigantic work force of replaceable men. The pyramids are the first monumental work of technology, although done with the crude hardware of ropes and pulleys. The "software" technique of human management was a demonstration of sophisticated technology!

According to this level of abstraction, technology includes the concept of hierarchy, bureaucracy as a form of management, systems analysis (that is, taking all factors into account!), and the placing of supreme value on efficiency and results ("By their fruits you shall know them"). Technology thus is not confined to a highly "toolized" way of doing work, but includes a great complex of human attitudes, theories of progress, and ideas about what constitutes human good and how it can best be served. Whether the mechanized hardware shapes people's attitudes, or whether people's attitudes create the machinery, is a chicken-or-egg question, but at this point (still trying only to describe and to avoid judgments) we must see that *technology by its very nature tries to avoid value questions, yet itself becomes a value.* Ellul states it vividly:

There is an attractive notion which would apparently resolve all technical problems: that it is not

THE TECHNOLOGICAL SPIRIT TODAY

technique which is wrong, but the use men make of it. Consequently, if the use is changed, there will no longer be any objection to the technique. . . . But all this is an error. It resolutely refuses to recognize technical reality. It supposes, to begin with, that all men orient technique in a given direction for moral, and consequently non-technical, reasons. But a principal characteristic of technique . . . is its refusal to tolerate moral judgments. It is absolutely independent of them and eliminates them from its domain. Technique never observes the distinction between moral and immoral use. It tends, on the contrary, to create a completely independent morality.[9]

Whether this characteristic of technology deserves a Christian Yea or Nay remains to be seen, but I am convinced that Ellul has stated as a fact what is a fact, namely that technology itself, by its own nature, refuses to abide by any values except its own, and its values are efficiency and material abundance and control of the environment through the systematic use of knowledge and tools.

Perhaps we have described and defined long enough and are ready now to move into a more careful probing of three cases where technology makes penetrating impact on our daily lives.

---

[9] Ellul, *The Technological Society*, pp. 96-97.

3

Creatures in the World of Nature

Once upon a time Americans who set out for foreign travel were warned against the drinking water, but nowadays that is changed. Instead, Americans must warn those who visit their shores:

> If you visit American city
> You will find it very pretty.
> Just two things of which you must beware:
> Don't drink the water and don't breathe the air.[1]

Tom Lehrer's song entitled "Pollution" ridicules our "hot and cold running crud," and brushing our teeth in "industrial waste," but it may still be safe to breathe "if you don't inhale."

The Abuse of Nature

All this damage to nature is no joking matter. In New York, thirty percent of the days during the first seven months of 1970 were declared "unhealthy" due to polluted air. In Tokyo traffic

[1] From *That Was the Year That Was* (Westbury, N.Y.: Cimino Publications, Inc.). Used by permission of Tom Lehrer.

CREATURES IN THE WORLD OF NATURE

policemen must return to headquarters every few hours to inhale oxygen, and in the coffee shops a customer can whiff one minute of pure oxygen for a quarter. On smog-alert days school pupils wear gauze-masks in class, and so many legendary cherry trees have died of bad atmosphere that Tokyo has asked Washington for saplings from the very trees Japan gave to the United States fifty years ago. One cynic remarked, "Things are not that bad here in America; you can still cut the smog with a knife!" California school children, however, are excused from outdoor games on days when the atmosphere chokes their lungs. Indeed, automobile exhausts and belching smokestacks pour 160 million tons of pollutants into the skies every year, and that warms the atmosphere, threatens to melt the icecaps, and wraps the cities with a thick blanket of dirt and disease.

The water may be worse. In Cleveland the Cuyahoga River is so polluted it recently caught fire, and someone said of the Hudson River that if a person fell in he would not drown, but would soon decompose. When factories and sewage plants dump their waste products, and the family washing machines and kitchen sinks pour their detergents into the nearby rivers, it makes a filthy system that kills the fish, prohibits boating and swimming, and makes an ugly scene.

Another contaminator of the natural world is

the chemical bombardment. For instance, DDT was first used to eradicate lice and control typhus, but when it was sprayed on the beaches it was deadly to fish, and by eating contaminated fish the falcons and osprey died. Used as a pesticide to control Dutch elm disease, DDT on the leaves fell to the ground, was absorbed into the soil and eaten by worms, which in turn became the poisoned food for robins which died by the millions. DDT decreases the ability of small animals to learn and remember, and some geneticists believe it damages the genetic health of animals and, most likely of humans. Its damage accumulates from small doses. DDT scatters everywhere because it decomposes very slowly, so it is found in penguins and seals in the Arctic regions thousands of miles away from its nearest use. It gets washed into the seas and threatens to kill off plankton that supplies 70% of the earth's oxygen. No wonder the Department of Agriculture banned DDT from general use by the end of 1970. Yet Secretary Finch warned that even then it would take ten years to purge the environment.[2]

If DDT were abolished completely, other chemicals remain, including the phosphates and nitrates used as fertilizers. A frightening excess of nitrate shows up even in the natural rainfall in

[2] As of late summer 1970, the political hassle continues. Several states have differing laws, and huge amounts of DDT are still sold, much of it abroad.

CREATURES IN THE WORLD OF NATURE

the midwest cornbelt, and major rivers of the country are overloaded with fertilizer drainage, increasing the growth of algae and destroying freshwater fish. Nitrate pollution is aggravated by automobiles, because car exhaust contains nitrogen oxides and rain carries them into the rivers. So every time a person drives his car, he adds to the spoilage of the earth's water supply.

When Pete Conrad set foot on the moon and let his camera scan the site, he blurted out, "Where is the earth? Oh, there it is!" That view from the moon gave a fresh glimpse into the truth of the matter: there is the earth, out there in the distant sky, "small and blue and beautiful," a "tiny craft in the enormous, empty night," a spaceship hurtling through space at a speed 600 times faster than a jet plane, carrying limited resources for sustaining its human cargo of 3.5 billion people. What is it able to provide to support that life? Above those people is a thin band of usable atmosphere, not over seven miles of it, with no "new air" beyond that. Beneath them, a thin crust of soil, only one eighth of its surface usable for food growing, perilously thin and easily eroded into the seas. Around them a finite supply of usable water which they must endlessly clean and reuse. That's it, and that's all there is, and there isn't any more.

We human beings travel and exist, we are born and we will die on this one spaceship Earth, and

PLENTY AND TROUBLE

we can preserve our lives only by "the care, the work, and the love we give our fragile craft." (Adlai Stevenson) "We are wedded to the good earth, and any damage to it means damage to our only home."[3] When we strip off the topsoil to expose the mineral deposits for easy profits, when we bulldoze out the forests and leave the hillside denuded, when we turn streams into open sewers, when we cut highways through the valleys and fill the air with pollutants that blacken the sky and blot out the sun—then we corrupt our own home and make ugly what is meant to be beautiful, and so we corrode the quality of human life.[4]

Unrecognized Assumptions Behind the Abuse

We are at best thoughtless, at worst ruthless. Industry uses the public rivers for the disposal of its private sewage. Schools and other public agencies belch smoke into the air. Farmers contaminate the soil with short-run pesticides and weed con-

[3] See René Dubos, *So Human an Animal* (New York: Scribners, 1968), pp. 144-45, for an eloquent statement that man "is earthbound forever because his life is completely dependent on fresh water and especially on the earth's atmosphere. . . . There is no hope whatever that man's biological nature can be changed enough to enable him to survive without the earth's atmosphere."

[4] Sketches of the eco-catastrophe recur in the popular press so often that it seems unnecessary to extend this one here. See, for instance, Frederick Elder, *Crisis in Eden* (Nashville: Abingdon Press, 1970), pp. 108-28.

CREATURES IN THE WORLD OF NATURE

trol. Everyone insists on his own automobile, even when he knows it pumps lead and carbon dioxide and other poisonous fumes into the air. Everyone seems impatient and self-concerned. Dubos comments, "Textbooks damn Louis XV for his irresponsible remark, 'After me, the deluge.' Yet we, too, are using the earth as if we were the last generation. Socially, we behave as if we were willing to excuse our misdeeds with the question, 'What has posterity done for me?' "[5]

The raping of nature is done by all of us, by our thoughtless and ruthless acts of violence against nature. We appear to assume that people have first class citizenship in creation. "Animals have no souls," Aldous Huxley reminds us,

> therefore, according to the most authoritative Christian theologians, they may be treated as though they were things. The truth, as we are now beginning to realize, is that even things ought not to be treated as *mere* things. They should be treated as though they were parts of a vast living organism. "Do as you would be done by." The Golden Rule applies to our dealings with nature no less than to our dealings with our fellowmen. If we hope to be well treated by nature, we must stop talking about "mere things" and start treating our planet with intelligence and consideration.[6]

[5] Dubos, *So Human an Animal*, pp. 184-85.
[6] "The Politics of Population," *The Center Magazine* II, no. 2 (March, 1969): 19.

PLENTY AND TROUBLE

The Unhappy Christian Contribution to the Abuse of Nature

Sad to say, there are elements in the Christian tradition that have contributed to the arrogance which men assume over nature.

1. *Conquest.* Those words in Genesis, for instance, have caused lots of mischief. "God created men, male and female, and God said to them, 'Be fruitful and multiply, and fill the earth and subdue it; and have dominion. . . .' " (Gen. 1:27-28). There it is! "Have dominion . . . subdue. . . ." Men have used those words to justify the conquest of nature. They have stripped the land of its grasses and the forests of their redwoods; they have poisoned the oceans so the blue whales perish, and shot the buffalo almost to extinction, and tormented rats and guinea pigs in painful experiments, and tortured animals training them for games and shows—all under the excuse that men are ordered by their Creator to "hold dominion and subdue"!

2. *Desacralization.* One direct benefit of Christian thought has been "desacralization"—that is, making not-sacred. By biblical insight, nature is not sacred; gods do not reside in the trees and rocks and streams and animals. We do not worship the nature gods of the Canaanites. That funda-

mental insight of biblical religion means that men can stand off from nature, study it, understand it. That opens up the vast possibilities of science. Science cannot exist if natural things are felt to be divine, but the Christian understanding that God creates nature but does not himself identify with nature—that insight protects the scientist's right to study nature without any fear of divine disapproval.

This insight can be abused, however, because it tends to encourage men to use nature however they see fit. Freed from fear of the spirits residing in the land, men are tempted to tamper with the land and subdue it to their own will. If creation is not divine, it appears then to be a limitless supply of material available for human consumption, a kind of promised land made to be possessed! This attitude has come under bitter criticism by Prof. Lynn White, Jr., who charges that Christian thought has contributed immensely to the desecration of nature. "To a Christian a tree can be no more than a physical fact. The whole concept of the sacred grove is alien to Christianity and to the ethos of the West. For nearly two millennia Christian missionaries have been chopping down sacred groves which are idolatrous because they assume spirit in nature."[7]

[7] "The Historical Roots of our Ecological Crisis," *Science* 155, no. 3767 (March 10, 1967) p. 1206.

PLENTY AND TROUBLE

White goes on to say that "Christian theologians generally have been unsympathetic, even hostile, to the natural religions that prevail among agricultural and primitive peoples." Indeed, we do understand God to be transcendent, removed from nature as its creator. He is Maker of heaven and earth, as the creed says, and not confined in that heaven and earth. He breaks into the world of nature by the Incarnation, by coming into it from outside, to put it sharply. Unhappily, this infers that nature is separate from God, and therefore available for exploitation.

If God is thus separated from nature, men feel themselves also separated. If God is not "in" the redwood tree or the Grand Canyon, then men feel free to axe down the redwoods and dam up the Canyon. According to our beliefs, nature has no rights of its own, it belongs to us to hold and subdue, because no spirits reside in it. Thus, nature is no longer for us a thing to contemplate or admire, but to use and to transform. God expelled Adam and Eve from the Garden; now we, their descendants, have returned the compliment and have taken over the Garden ourselves to subdue it and hold dominion as conquerors. Harvey Cox describes the results: "Man's attitude toward disenchanted nature has sometimes shown elements of vindictiveness. Like a child suddenly released from parental constraints, he takes savage pride in smashing nature and brutalizing it."[8]

As a partial explanation for this ravaging attitude we do well to remember that man's struggle against nature arose out of his early experience with nature as a hostile enemy. Nature was cold and barren and it threatened to destroy him. This sense of being nature's victim still prevails among some people. Eric Hoffer, for instance, has written about it vividly.

All through adult life I have had a feeling of revulsion when told how nature aids and guides us, how like a stern mother she nudges and pushes man to fulfill her wise designs. As a migratory worker from the age of eighteen I knew nature as ill-disposed and inhospitable. . . . The truth about nature I found in the newspapers, in the almost daily reports of floods, fires, tornados, blizzards, hurricanes, typhoons, hailstorms, sandstorms, earthquakes, avalanches, eruptions, inundations, pests, plagues and famines. . . . It seemed to me we were surrounded by devouring, pitiless forces, that the earth was full of anger, the sky dark with wrath, and that man had built the city as a refuge from a hostile, nonhuman cosmos.[9]

I concur that much contemporary writing about nature and ecology infers that nature is pure, in-

[8] *The Secular City* (New York: Macmillan, 1965), p. 23.
[9] "A Strategy for the War with Nature," *Saturday Review,* Feb. 5, 1966, p. 27. Copyright 1966 *Saturday Review,* Inc.

PLENTY AND TROUBLE

nocent, serene, and, as Eric Hoffer goes on to say, a "health-giving, bountiful fountainhead of elevated thoughts and noble feelings. It seem(s) that every writer was a 'nature boy.'" I hold no such view. Christian honesty requires that we be realistic about nature. Nature is apples *and snakes*. Having said that, and holding that clearly in mind throughout the rest of our study, let us go ahead with our survey of what Christian thought has contributed to the abuse of nature.

3. *History over Nature.* Still another Christian concept that makes for the abuse of nature is the idea that men are redeemed in their social, not their biological experience; that is, in history, not in nature. What counts is that men shall "love God and their fellowmen." Yet that says nothing about their love for nature. It makes no confession that men are biological beings as well as social beings. It emphasizes their humanity and neglects their creaturehood. According to this understanding, nature is only the stage where the human drama is acted out. Emil Brunner once put it bluntly: "The cosmic element in the whole Bible is never anything more than the 'scenery' in which the history of mankind takes place."[10] That sounds scandalous when you reflect on it. At the

[10] For another statement of this idea, see Richard A. Norris, *God and World in Early Christian Thought* (New York: Seabury Press), 1965, p. 38.

CREATURES IN THE WORLD OF NATURE

very least it seems blind to half of human nature; at most it distorts and cripples our self-understanding. Another theologian describes this disaster:

Christian man has struggled to disengage himself emotionally from nature in the name of the transcendent God and on behalf of his immortal soul. If not his divorce, then his "legal separation" from nature was effected in order to concentrate energy upon a love of persons in accord with the love of Jesus Christ; but this emphasis drew from him the view that nature is a lifeless instrument of man's "spiritual" existence, and that man himself in his "fleshy" capacity represented a force running counter to the spirit. This attitude to nature showed itself first in asceticism of the body and then in the technological conquest of nature.[11]

4. *Matter is Evil.* This led to the heresy and perversion that spirit is good and matter evil. Christians forgot that people have bodies as often as they have souls. Bodyless souls are rather rare! "In the interests of 'spiritual' religion the natural world was regarded as an enemy. In Protestantism everything human and natural [here Bonifazi means everything fleshy and earthy] sank into the night of sin."[12] Salvation suggested soul, with

[11] Conrad Bonifazi, *A Theology of Things* (Philadelphia: Lippincott, 1967), pp. 158-59.
[12] Conrad Bonifazi, "Christianity and the World of Nature," *Humanity,* April, 1967, p. 4.

nothing to do with bodies or sex or disease. The physical body was considered a downdrag on the soul, a handicap to spirituality. All this was a perversion of the biblical view that God saves the whole man, not just the soul. Real blindness is cured, real diseased hands restored, real legs walk, real death is conquered. As Joseph Sittler put it, "God is the undeviating materialist. He likes material, he invented it."

In summary, it must be confessed that Christian thought has often contributed to the abuse of nature. The Yale theologian, Julian Hartt, makes bold to say that some Christian ideas have "legitimized man's total exploitation of his environment." And the theologian Pogo tells it the way it is, "We have met the enemy and he is us." One greater than these once warned, "Let him who is without sin among you cast the first stone." The chosen people once forgot, but were taught to remember that they were chosen for responsibility, not for privilege. So man, the creature, especially the Christian man, has forgotten but is now being taught to remember that he, too, is chosen by the Creator for responsibility, not for privilege.

A Proper Christian Understanding of Nature

1. *The Inalienable Rights of Nature.* There is another strand of Christian thought that gives

CREATURES IN THE WORLD OF NATURE

proper respect and love for the world of creation. Its most eloquent spokesman was, of course, Francis of Assisi, a devout troubadour, a joyful singer, a young man strikingly like a modern hippie in some ways and like his Master Jesus Christ in most every way. As everyone knows, Francis preached to the birds, tamed a wolf, kissed lepers, converted robbers, and married Lady Poverty. He had a love for immediate things, not just for the world in general; he loved men, not just mankind. He had regard for flowers, streams, bread, stones, as well as for beggars and the sick. He gave personal attention to natural things, endowing them with a kind of personality. Nowadays we consider such an attitude ridiculous, although we permit it to poets and mystics and preachers. For St. Francis every natural object had a significance of *its own;* it was not simply raw material for Jesus' parables about people. All created things were *intrinsically valuable in their own right.*[13]

A careful reading of some parts of the Bible

[13] A passage from a chaplain friend of mine reads: "I come to a halt and I contemplate a blossoming apple tree, for example. I do not immediately think about its cells or about the food it may produce for me, nor do I immediately think that the tree may need pruning or spraying. . . . The blossoming apple tree stands before me in its own right, a beautiful entity posited there for its own sake. I contemplate it and am captivated by it. I do not penetrate behind its sheer givenness." H. Paul Santmire, "I-Thou, I-It and I-Ens," *The Journal of Religion* 48, no. 3 (July, 1968) p. 268.

PLENTY AND TROUBLE

reveals the same thing. Natural things sing their praises and honor directly to God. They celebrate God's glory and majesty as spokesmen for themselves. "The heavens are telling the glory of God; and the firmament proclaims his handiwork." (Psalm 19:1) In turn for their praise, God cares for every creature. The magnificent passage in Job 38–39 details the careful attention God bestows upon each creature, one by one. The Psalmist declares how God cares for these for their own sake:

Thou makest springs gush forth in the valleys;
   they flow between the hills,
they give drink to every beast of the field;
   the wild asses quench their thirst.
By them the birds of the air have their habitation;
   they sing among the branches.
From thy lofty abode thou waterest the mountains;
   the earth is satisfied with the fruit of thy work. . . .
The high mountains are for the wild goats;
   the rocks are a refuge for the badgers.

                       (Psalm 104:10-13, 18)

There is no utilitarian word in all that. The creatures are not made for the benefit of man, but in their own right they exist and in the care of God they prosper. Every bit of creation has its intrinsic and inalienable rights. The blue whales have a right to the sea, the osprey have a right to healthy fish, the streams have a right to run clean and pure in their course, the air has a right to blow clean, and the redwood has a right to stand and to grow tall.

2. *The Restoration of All Nature.* According to the biblical understanding, all nature suffers from human wrongdoing.

There is no faithfulness or kindness,
  and no knowledge of God in the land;
There is swearing, lying, killing, stealing, and committing adultery;
  they break all bonds and murder follows murder.
Therefore the land mourns,
  and all who dwell in it languish,
and also the beasts of the field,
  and the birds of the air,
  and even the fish of the sea are taken away.
                              (Hosea 4:1-3)

Then in the Christian vision of the time when all creation will be fulfilled, men certainly will enter into the glory of God, and *all creatures also* will share in that redemption. "The mountains and the hills before you shall break forth into singing, and all the trees of the field shall clap their hands." (Isaiah 55:12; see 43:19-20) Meanwhile the whole creation "waits with eager longing for the revealing of the sons of God; for the creation was sub-

PLENTY AND TROUBLE

jected to futility, not of its own will but by the will of him who subjected it in hope; because the creation itself will be set free from its bondage to decay and obtain the glorious liberty of the children of God. We know that the whole creation has been groaning in travail together until now." (Romans 8:19-22) We cannot be sure what all St. Paul had in mind in this passage, but Christian people share the hope that not only their own life, but the life of birds and creeping things, the trees and the mountains and the living air will also share in the restoration when things become what God means them to be. In many passages there seems to be something of the "Franciscan awareness of rapport." (Moule)

3. *Take Care of the Garden.* To prepare for that day people should pay attention to that other verse in Genesis, "The Lord God took the man and put him in the garden of Eden to till it and keep it." [14] Men are to dress the earth, and tend it, and make a garden of it, and treat it with the respect a guest shows in the home of his host. They are to enjoy it, and use it, and not abstain from using it, but they are to care for it, and render account to God for their use of it, for the earth is God's footstool. They remember that *it is God* who put man in the Garden, and it is

[14] Genesis 2:15, from the earlier, primitive, "J" account of creation.

CREATURES IN THE WORLD OF NATURE

God who made man a little *lower than God* (Psalm 8:5), so he is nature's servant, not its lord and master. Tagore once called this the "active wooing of the earth." Albert Schweitzer, as a boy, refused to shoot birds. He had sympathy for an old, overworked asthmatic horse. One day when he cracked a whip at a dog barking at his carriage, he caught the dog's face and it went howling off the road. Then "I heard its cries for weeks. I vowed I would never let my feelings get blunted, or be afraid of the reproach of sentimentalism." Schweitzer was unashamed to be kind; he had a reverence for the life in every creature.

We need what Scott Paradise calls "a new materialism"—that is, a proper concern for the material world, the whole, living, organic creation. This "new materialism" would require a covenant between man and nature to correlate with "the everlasting covenant between God and every living creature" (Genesis 9:15, 16) which God established after the flood and made the rainbow to symbolize. God asks people to cherish and embrace the organic world. Much as he instructed Noah to save every living species in time of natural disaster, so he instructs modern men to save the creation from human disaster. God created a finite world. "Finite" means that resources for human living are limited, interrelated, and destructible. Limited: oxygen and water can be exhausted; so

PLENTY AND TROUBLE

can forests and fresh air. Interrelated: burning of fossil fuels could fill the air with carbon dioxide and make an air blanket to smother the earth into a giant hothouse, melting the icecaps; or, by filtering out the sun's rays, the cover might turn the planet into a giant refrigerator. (Man would be undeservedly lucky if these two trends cancel each other out!) And destructible. Nuclear explosions have already blasted whole islands out of the sea. Therefore, finite men must learn to care for the finite creation. Men have no right to poison the world, nor to proliferate their species to the point of suicide. They have the clear duty to remember that God became incarnate in the natural world of flesh and blood and used the natural elements of grape and grain because he loves the natural world. Pollution, therefore, may for the secular man be senseless, but for the man of faith it is blasphemous. It profanes what God creates and loves. But the Flood (in some form) still threatens disobedient people. As Paul Ehrlich says it, Nature bats last!

4. *People Belong to Nature.* The proper Christian understanding recognizes also that man is himself part of nature. He belongs among the creatures. Biologically this is as obvious as Darwin. Religiously it means that the Catholic writer Michael Novak is right:

CREATURES IN THE WORLD OF NATURE

Man is not, in the phrase of Alan Watts, an ego trapped in a bag of skin, alien to his environment. Man is a part of nature, brought forth from the universe like fruit from a tree. The universe is a *thou* to him, inseparable from his own self, part of him, and he is in dialogue with it. As an apple tree *apples*, so the universe *peoples*....[15]

Is there not something of rare beauty in this sense that human beings belong to nature as an apple belongs to the apple tree? Perhaps this is what Teilhard de Chardin has been teaching, that the universe is a single web of natural and human history, growing ever more complex and ever more self-aware. And what George G. Simpson, distinguished vertebrate paleontologist, means:

[Man] is in the fullest sense a part of nature and not apart from it. He is akin, not figuratively but literally, to every living thing, be it an amoeba, a tapeworm, a flea, a seaweed, an oak tree, or a monkey. . . . This is togetherness and brotherhood with a vengeance, beyond the wildest dreams of copy writers or of theologians.[16]

If man is a part of nature, and if Buber's I-Thou, I-It distinction is meaningful, then we are

[15] *A Theology for Radical Politics* (New York: Herder and Herder, 1969), p. 96.
[16] *This View of Life* (New York: Harcourt Brace & World, 1964), pp. 12-13.

prompted to say that human response to nature ought to regard nature more as a *thou* than an *it*. That would create all kinds of theological trouble, however, as H. Paul Santmire warns. When he explored this question, he came to "a better alternative," and he calls nature neither *Thou* nor *It*, but *Ens*.[17] The Ens exhibits a certain mysterious activity. I cannot fully predict how the Ens will behave; it does not match my expectations. (The weather forecasters know this for sure!) It will grow, decay, develop, stand still, or disappear in ways I cannot fully understand. (Doctors know this!) I give Ens my total attention in a mood of wonder and humility. I am repulsed by its dark side, as by Moby Dick, and I am delighted that "God has stretched out his hand to give us the splendor of the sun and the moon to enjoy," as Calvin said. As both Melville and Calvin make clear, nature arouses in me the sense for the presence of Deity. Nature itself seems to be open to the Infinite, and because I, a man, a human creature, am also open to the Infinite, I belong to nature.

Let us beware, however, of putting a halo on Mother Nature. "She" is not the Virgin Angel. At times cruel, "red in tooth and claw," competitive and violent, with the strong preying on the

[17] "I-Thou, I-It, and I-Ens," pp. 260-73. Here I paraphrase and condense his wording, and intersperse comments of my own.

weak, nature includes predators and parasites. If we baptize nature with sacramental meaning, we invite the moral disaster that nature becomes the model for morality. That would justify aggression. The ruthless would dominate by their cunning and combat, and technological man would rule all the kingdom of nature with moral impunity. He would be lord and lion of the jungle.

5. *Summary.* In our dealings with nature, then, we are not to worship nature, nor treat her as a slave or a machine. Instead, we stand alongside nature as a cherishing brother, for nature, too, is God's creation and bears his image. We are neither to adore nature nor plunder her, but tend and care for her. Behind the naive images found in Psalm 104, a person can locate profound insights into the majesty and wonder of creation. That psalm declares that there is nothing in the world of man or nature that is independent or solitary; it all belongs together. Every small thing depends upon all others, and every item reflects something beyond itself. Light is the garment God wears, the heavens are the curtain for his dwelling. The heavy voice of thunder is his rebuke, and the rains shower his generosity upon the beasts. Trees and birds, grass and cattle, and the vine that gladdens the heart of man are all bound together in the bundle of God's grace. "Natural and mortal

PLENTY AND TROUBLE

life are incandescent with meaning because they all depend upon the will of the ultimate and Holy One." (Sittler)

These all look to thee,
   to give them their food in due season.
When thou givest to them, they gather it up;
   when thou openest thy hand, they are filled with
      good things.
When thou hidest thy face, they are dismayed;
   when thou takest away their breath, they die
   and return to their dust.
When thou sendest forth thy Spirit, they are created;
   and thou renewest the face of the ground.
                            (Psalm 104:27-30)

What has technology done to change any of the essential wisdom of such lines? Is it not just as possible today as yesterday, and even more necessary now than then, to see that water is not water only, and still less, a substance for transporting our sewage and waste, but that water is intended to cleanse and to refresh; as cleansing it suggests our baptism, as refreshment it recalls that "he leads me by the still waters, he restores my soul." Likewise the hills are not hills only; they are more than stone for building and wood for furniture; they are a shoulder against the storm, and "thy righteousness is like a great mountain." The living air is not a mixture of gases only, something for the chemist to separate, but air is a likeness of the invisible Presence that blows where it will. "Thou, O Lord, dost ride the wings of the wind, and make the winds thy messengers, fire and flame thy ministers." (Psalm 104:3, 4) Jesus said the sky is God's throne, the earth his footstool, the lost sheep the parable of our human lostness, and the growing seed a sign of God's renewing of our world. "Sad is the modern man who lets his technology deprive him of the truth hidden in such words." (Sittler)

Surely in the plan behind all things, we human people are meant to care for the creation, to use it, nurture it, redeem it, but not to tyrannize over it, for we, too, are creatures; nor to neglect it, for we are more than creatures and are given dominion over all creation. We share the image of the Creator in a way the rest of creation does not; hence, we are responsible for it, as Adam was made responsible. As soon as we begin to use it selfishly and reach out to take fruit which is forbidden, "instantly the ecological balance is upset and nature begins to groan." [18]

---

[18] C. F. D. Moule, *Man and Nature in the New Testament* (Philadelphia: Fortress Press, 1964), p. 14.

4

What Is Man More Than a Machine?

As we said at the beginning, men are tempted to interpret the past in terms of what interests them in the present. Being devoted to material goods, modern men tend to see their history as a struggle for material goods. Taking delight in mechanical tools, they tend to regard their ancestors as men who worked with tools; they think of Adam as *homo faber*, toolmaker. Many people therefore, consider toolmaking the fundamental characteristic of human nature.

The Nature of Human Nature

There are options however. Prof. Huizinga believes that play, not work, is the formative element in human culture, and a man's most precious talent is not his serious activity, but his capacity to make-believe. Hence, man is *homo ludens.* Harvey Cox reminds us that a man sings, dances, prays, tells

WHAT IS MAN MORE THAN A MACHINE?

stories and celebrates; he is *homo festivus*. He is also a visionary dreamer and mythmaker, hence *homo fantasia*.

Man is *homo festivus* and *homo fantasia*. . . . But in recent centuries something has happened that has undercut man's capacity for festivity and fantasy. In Western civilization we have placed an enormous emphasis on man as worker (Luther and Marx) and man as thinker (Aquinas and Descartes). . . . This worker-thinker emphasis . . . helped produce the monumental achievement of Western science and industrial technology.[1]

Cox bewails the effect of this, saying something joyful and distinctly human has been lost.

Lewis Mumford regrets that Western thinkers tend to define man as a toolmaker. He points out that man certainly had speech, rituals, and social organization before he made tools; and what man possessed in singular degree was mind; he was *homo sapiens*.

I submit that at every stage man's inventions and transformations were less for the purpose of increasing food supply or controlling nature than for utilizing his own immense organic resources and expressing his latent potentialities, in order to fulfill more

[1] "In Praise of Festivity," *Saturday Review*, October 25, 1969, p. 26.

adequately his superorganic demands and aspirations.[2]

Mumford goes on to point out that the discipline of toolmaking and tool-using was a timely correction to the extravagant powers on invention that resulted from language, but "the main business of man was his own self-transformation." He was *homo sapiens,* a whole being who had the unique capacity to shape together into one imaginative personality all his powers of speech, ritual, play, fantasy, social organization, *and* toolmaking. "His biggest find and his first shapable artifact was himself."

Now it is within that context that I want to point out that man is *homo faber in order to be homo sapiens.* He makes tools for the purpose of being human. In order to think and to function as a distinctly human being, a man devises machines. And the machine *par excellence* is the computer.

How the Computer Grew Up

Ever since our first conscious moments, we human beings have been plagued by our limitations. Set against our problems and our enemies we were weak, and we wanted to break out from

[2] *The Myth of the Machine,* p. 8.

WHAT IS MAN MORE THAN A MACHINE?

our limitations and be supermen. Therefore, we devised tools and machines to supplement and extend our human powers. The story from the stone wheel to the Apollo spacecraft is an endless saga of our struggle to become stronger, faster, and more skilled. First we extended our physical powers and developed simple tools to lift logs, to plow fields, to whittle, and to bake. One of the climactic machines of this kind is the Mosher-monster,[3] a robot device, a 3,000 pound, 11-foot-high walking machine that lumbers along at five miles an hour, lifts 500-pound loads with ease, and kicks aside a 175-pound beam as though it were a matchstick. Thus do robot devices multiply the strength of a man's arms and legs. Next we extended our sensory powers and devised the glass lens, the telephone, the radio, and delicate instruments to amplify our powers of sight and hearing and touch. Now in our time, a third extension is working to expand our cognitive powers, to intensify our ability to think and reason and learn because here, too, we are limited. Except for a few geniuses such as Leonardo and Edison, we seem unable to grasp all the relevant material, to retain it, then recall it, sort it out, and relate it to our problem at hand. Yet that is precisely what

[3] Nicknamed for Ralph S. Mosher, inventor and developer in the G.E. Laboratory. Officially it was named CAM, the "Cybernetic Anthropomorphous Machine."

PLENTY AND TROUBLE

the computer does, in ways that astound us. This new technology magnifies our mental powers.[4]

It all began with counting. Prehistoric men used their ten fingers and a collection of twigs and pebbles. The Chinese devised the abacus, based on fives and twos. In 1623, a professor of biblical languages and astronomy named Wilhelm Schickard designed a machine to add and subtract, but it was destroyed by fire—along with his lecture notes! A few years later Pascal invented the first true calculating machine to survive, and in 1801 a French weaver used punched cards to control the patterns woven on his loom. In 1833, an Englishman named Charles Babbage thought up the first digital computer, aided by Lady Lovelace, who invented the language which is now the basic language of all computers, namely zero and one, on and off, yes and no. In our own time, at Harvard University a Ph.D. student hooked to-

---

[4] "The purpose of the computer is to enable us not to spend time on 'controls,' but to use time for tasks that require perception, imagination, human relations, and creativity." Peter F. Drucker, *The Age of Discontinuity* (New York: Harper & Row, 1969), p. 259. "Machines will, one hopes, bring about such a vision of the task of humane scholarship: to seek wisdom. . . . By freeing man from petty tasks that weary his mind, they will probably make triviality obsolete. . . . The machines will certainly not make genius obsolete nor put thinkers out of work. They will help us to see, however, that thinking is not to be reserved for technicians and professionals." Jacob Neusner, "Scholars and Machines," *The Christian Scholar* (Winter, 1960) XLIII: 273.

WHAT IS MAN MORE THAN A MACHINE?

gether some 78 adding machines and desk calculators under control by a paper tape punched with holes. "Mark I" could do three additions per second, but it is already obsolete and is displayed in the Smithsonian Museum as a "relic" of the early computer age. Then came electronic computers. The first one weighed 30 tons and occupied 1500 square feet; like a pro football lineman it was big but fast and could do 4500 additions per second. When the bulky vacuum tube was replaced by the transistor, the computer became much smaller and faster, and upped its performance to 204,000 additions per minute. Further microminiaturization compressed the whole works into a tiny chip one eighth of an inch square, containing 72 circuits complete with transistors, resistors, diodes, and wiring. A computer can now perform 15 million additions in one second; its operations are measured in nano seconds; one nano second is the slice of a second equal to the fraction of one second to thirty years.[5]

The Amazing Powers of the Computer

The one hundred thousand or more computers already in use in this country vastly extend the

[5] This historical sketch was taken from an IBM supplement in the New York Times.

PLENTY AND TROUBLE

human capacity to think, calculate, measure, and predict. Let your mind skip down a list of things already done or soon to be done by the computer. Banks, of course, use computers to run checks through the customer's account and to type out his notices of "insufficient funds." Industry uses computers to make out paychecks, keep the daily inventory, and calculate the times for repairs and replacement of machinery. Meteorologists use the computer to predict the weather; it does not improve the weather. In agriculture, sensing devices feed the computer information about soil moisture and chemical analysis, the temperature and long-range weather, then the computer calculates the best seed to plant, when to plant, the kind and amount of fertilizer, and when to harvest. The computer has not learned how to milk the cows.

For health, the computer records a person's physical history while he answers questions, and it analyzes his bodily functions as he sits before a screening device. It can give the doctor suggestions for treatment based on the patient's history and symptoms. During an operation the computer monitors his internal condition and gives the surgeon instant information.

For government, the computer already collects financial information on all citizens from their W-2 forms, their honorarium payments, gambling profits, and such; that is supposed to keep them

WHAT IS MAN MORE THAN A MACHINE?

honest! It has been proposed that a huge data center gather dossiers on all citizens, including the census information, court records, traffic offenses, and such "innocent" information as his civic clubs, marital status, and personal tastes and reading habits.

Already Manhattan has experimented with a computer control on traffic offenses. A patrolman at the entrance to a bridge radios to a computer the license numbers of passing cars. The computer is programmed with the license numbers of 30,000 stolen cars and 80,000 unpaid-traffic-ticket-holders. If the passing car is one of the wanted ones, the information is radioed to a police car at the bridge exit and the driver must pay on the spot—with his credit card, of course! Traffic will be controlled by the computer. In the future the highways may be so crowded a person will have to get a reservation to drive, as he now gets an airplane reservation by computer, and his car travel will be guided just as airplane travel now is guided by computer calculations of safety.

Scientists used the computer to calculate the movements of stars and planets for centuries back, and thus they figured out the purpose of those huge stones erected in a pattern at Stonehenge, England, some three thousand years before Christ. They were themselves a kind of computer, used to record and predict the movements of the stars and planets!

Libraries will use computers to search the data files for any specified material, and to print out a copy of some article a person asks for. And, of course, to keep track of borrowed books and to assess fines. Computers are the bill collectors of the future!

The computer will contribute to the teaching process in the classroom. In a recent letter a friend of mine in Milwaukee noted that she had the new experience of teaching by tape, telelecture, and telephone to classes in Green Bay, 150 miles away. Some classrooms are now equipped with individual consoles for each pupil, giving the teacher four devices to work through: a television screen, a typewriter, a movielike screen, and a headset through which the child gets spoken commands and suggestions. In turn, the child answers the computer in three ways: he can write with a pen that casts a beam of light onto the cathode tube behind the TV screen, or speak into a microphone, or type his reply onto the teletypewriter. Such teaching machines can personalize the instructions to fit the unique needs of each pupil. In the future, self-instruction in science, art, history, and languages may well become a consuming hobby.

The home of the future will include a computer console, tied into a central community computer,

WHAT IS MAN MORE THAN A MACHINE?

and citizens will rent this service as they now pay for telephone service—using their credit cards, of course, for money will be abolished except for small change. They can then order any TV program, past or current, and have it projected on their home screen. This console will open up a new range of experiences. There a person will do his voting, send messages to his Congressmen, and do his grocery shopping. Some men will be able to do their business from the console, and to hold conferences with colleagues. That will abolish many committee meetings, even church meetings. Imagine prayer meeting as a non-meeting! "Where two or three are tuned in, there? . . ."

There is another type of computer, the analogue computer, distinguished from the digital computer. The analogue computer can read handwriting, and that will sort out the mail. It already sorts 43,000 letters per hour if the address is typed, with zip code. This computer can read designs and drawings, and can transform a rough hand-sketched design into exact specifications. Then it reproduces such figures at a distance. For example, a computer "painted" a large portrait of Mona Lisa from a 35 mm color slide; a scanning device reduced the picture to 99,840 numbers, which were printed on a magnetic drum memory, then projected onto a screen miles away by translating the numbers into light and color values.

PLENTY AND TROUBLE

When Webster's dictionary, or the sounds of the piano keyboard, are fed into this computer and it is instructed to choose all possible patterns within a given definition of what is wanted, it will compose poetry, or sonatas.

Perhaps the most spectacular performance we all saw on TV, relayed via satellite. Apollo 12, after the moon landing, splashed down in mid-Pacific within a few seconds and a few yards of the moment and spot calculated for it months before. Such astounding precision of work has more than fulfilled the wildest dreams of men who want to extend their intellectual powers. The computer is so fast, so accurate, so comprehensive in its grasp of all the relevant material, that the old joke takes on fearsome point. The operator asked the computer, "Is there a God?" The machine whirred and coughed and stammered out, "There is now." It seems fitting to nail on the wall of the computer room the motto, "To err is unlikely, to forgive is unnecessary." One man who had an argument with the computer about how much he owed the department store suggests that the motto should read, "To err is human; to really foul things up requires a computer."

The Fourth Blow to the Ego

I have extolled the achievements of the computer to such length that you are bored by it, and I

WHAT IS MAN MORE THAN A MACHINE?

did so with a purpose. I want you to have the feeling, by reading a long list of the computer's capacities, that you are overwhelmed by it—exactly as men feel who work on these machines. The computer works day and night without a coffee break; it produces endlessly, accurately, with incredible speed. The computer adds one final blow to the human ego.

There have been three serious blows before now. (1) Copernicus showed that the earth, our earth, our home, is not the center of creation, but a mere speck of dust whirling around a third rate sun, far off in one corner of the cosmos. (2) Then Darwin made it clear that human life runs on a continuum with other animal life, which emerged out of earlier, simple forms of life, back, back to amoeba and tiny cells of living stuff. We humans are not a once-and-for-all sudden creation. (3) Then for the third blow came Freud, revealing that no man is master in his own household. Irrational elements of the id and the unconscious hold great power over his thought and behavior. (4) Now finally comes the computer, the machine, the most damaging of these blows, doing work with speed and precision and breadth of grasp no man can match.[6]

[6] See Bruce Mazlish, "The Fourth Discontinuity," *Technology and Culture* (January, 1967), VIII: 1-15.

PLENTY AND TROUBLE

When a man is told he works like a machine he may consider it a compliment momentarily,[7] but usually he feels affronted. The machine threatens his last toehold on dignity because it excels him. It feels superhuman, and compared to it he feels himself to be a feeble worker, slow, disorganized, and restricted.

Even when he plays games with the computer, it beats him! In chess the computer wins over its programmer because it never repeats a mistake; it learns, remembers, and even "takes into account the personality of the opponent," remaking its own strategy to beat the style of its competitor.[8] The machine seems to take on personality. A cartoon shows a schoolboy giving his report card to his father and saying, "My teaching machine hates me." And the machine may even converse! I heard a machine talk. A TV newsclip showed a machine that helps the blind to read. With a scanning device it picks up the printed image on any page, transfers each letter via 144 tiny circuits, then transcribes these into words, then into voice impulses, and the machine audibly reads the page! This machine speaks. It is almost human! What

---

[7] Certainly the Baltimore baseball coach meant it so when he said about his best relief pitcher, "He's a human computer. He has in his head a file on all batters, and he never forgets."

[8] Harold E. Hatt, *Cybernetics and the Image of Man* (Nashville: Abingdon Press, 1968), p. 157.

is left that a man can do to hold any superiority over the machine? It feels alive.[9]

Machines Almost Human

The electronic computer bears some resemblance to the human brain, and vice versa. The human nervous system functions as electrical impulses pass through nerve cells (that is, neurons) and the brain is a vast network of neurons in elaborate structures with extremely complex interconnections. The neuron "fires" or it does not "fire"; it is "on" or "off," "yes" or "no"—just as the transistor signals on or off, one or zero—and that simple signal is then transferred and translated into meaningful output messages from the computer and from the brain. This means the human brain is analogous to the computer, similar in operation, though not identical in function.[10] The computer handles only quantitative data, such material as can be described in mathematical

[9] A rising flood of articles and books testifies to the anxiety about this. The best book I know is the above by Harold E. Hatt, who remarks on this matter: "The 'takeover' of our world by machines has quite astounded us, and I think that at least some of the humor about it is a nervous humor, as, for example, the remark that one of the most electronic computers is so human that it blames its mistakes on other [computers]," p. 151.

[10] See Norman J. Faramelli, *Some Implications of Cybernetics for our Understanding of Man,* an unpublished dissertation, 1967, espec. pp. 99-100.

or logical terms: finite, exact, discrete material. It cannot handle loaded words such as "beauty," "honor," "trash," "lie"—so you have to watch your language when you talk to the computer! But the point is that "the human brain functions in a manner analogous to the electronic computing machine."

The distinctive operating principle of cybernetics (the use of computers for control and communication) and the principle which corrects mistakes in thoughtful human life both demonstate the principle of feedback. A thermostat provides the classic illustration of feedback. When the temperature goes down to 68, the thermostat sends a message to the furnace, "Throw the switch and turn on the gas." When the room gets to 72, the thermostat sends another message, "That's enough; turn off the gas." By means of the thermostat, information about the furnace's performance is fed back to the regulating device to correct the performance and bring performance into accord with the intention.[11]

In a similar way, the human body has feedback devices. A man goes duck hunting and his first shot misses five feet to the south; his eye sends a message to the brain to tell the arm to slow down a bit, and fire again; this time he misses it five

---

[11] Norbert Weiner, *The Human Use of Human Beings* (Garden City, N.Y.: Doubleday, 1950), p. 27.

feet to the north. Of course, the duck falls dead, killed by the law of averages! The moral of this story is that the thermostat does more accurate reporting than human eyes can do—but both the thermostat and the eye are feedback devices. The human body has many internal feedback mechanisms. When a person gets a chill, that stimulates the brain and he shivers; that develops heat to counteract the chill. If he has a hemorrhage, that drops the blood pressure and the arteries contract; that cuts the blood flow and eases the hemorrhage. The human body has a large "assembly of thermostats, governors and the like." (Weiner)

There is such a similarity between the computer and the human brain that it gets hard to find a clear distinction between the living and the nonliving. Let us try to understand the matter by considering the manlike characteristics of the machine.

Can the machine *remember?* The computer certainly remembers information that is stored in cards or tapes, and it can retrieve it almost instantaneously; but how human beings remember is still a puzzle to the biologists. We *do* remember, but just *how*, no one can say.

Do machines *learn?* It depends on the definition. If by learning we mean "being made able to do" a certain thing, such as being made able

PLENTY AND TROUBLE

to read or skate, then the computer certainly does learn; it learns to play checkers, and it learns from the experience of playing checkers, to the point it can beat the programmer who teaches it.[12] Learning from one's experience is something lots of people don't do, including drunks and nations drunk on visions of grandeur.

Can machines *think?* Again, it depends on definition. If thinking means something inscrutable and mysterious, then machines do not think; there is nothing mysterious about what they do. But thinking can be defined as the mastery of a certain technique. Listen to Norman Faramelli describe it:

A machine can be said to be thinking if it can perform some logical operation which, if done by a human being, would be considered intelligent. That is, if a particular action performed by a human being deserves to be called an intelligent action, then a machine that performs the same operation can be said to be functioning intelligently, or is thinking. It must be kept in mind that we are dealing with a functional definition only. Thinking is viewed in behavioral terms; it is a tool.[13]

Obviously the machine cannot do all the thinking that a human can do; but the Wright Brother's

[12] Faramelli, *Some Implications of Cybernetics,* pp. 213-14.
[13] *Ibid.*, p. 215.

WHAT IS MAN MORE THAN A MACHINE?

contraption at Kitty Hawk could not do all the 747 can do, yet who will deny it was an airplane? The computer can analyze performance, diagnose trouble, and make changes. It can think as well as middle-level people who "use their minds." (Donald N. Michael) They are replaced by computers! "Spontaneity, novelty, and creativity were once thought to be exclusively human phenomena. But these qualities can no longer be denied to machines. Machines have made unexpected moves in games, have discovered original proofs in theorems, have composed works of art, and so forth."[14] Sir Leon Bagrit, in his BBC Reith Lectures in 1964 said, "It is possible to build computers with something approaching the amazing flexibility of the human brain." And Prof. Hatt concludes his study of this subject with the conviction that "we have acknowledged our kinship with the animals. Now we need to concede that some machines are closer to us than are most animals, and perhaps than all animals."[15]

Listen again to Faramelli:

The work that is currently being done in artificial intelligence is an attempt to push machine behavior

---

[14] Hatt, *Cybernetics and the Image of Man*, p. 175.

[15] *Ibid.*, p. 202. See also Norbert Weiner, *The Human Use of Human Beings*. Chapter 3 deals with the learning capacity of machines, and chapter 4 with their linguistic capability, that exceeds that of the chimpanzees although chimpanzees are among "man's closest relatives and his most active imitators."

PLENTY AND TROUBLE

further out into (the continuum of speed and complexity in the processing of information). Does this mean that there is no boundary between what a machine can do and what a man does? Of course not. Current study does not indicate that there is or is not an impassable boundary. It merely states that if there is an impassable boundary, it is much further out than is the case of today's computers.[16]

Certainly we ought not by definition set any *a priori* limits to what a machine can do. Let's wait and see.

Elting E. Morison makes the bold assertion that

Computers . . . can discover proofs for theorems of logic; they can solve trigonometric identities; they can do formal integration and differentiation. They can already, or have given clear indication that they will be able to in the foreseeable future, do the following things: remember, learn, discern patterns in loose data, make novel combinations of old information, and, most striking of all, introduce surprise into an intellectual situation. There is still a question about the limits of the computer's imagination—its ability to create something like a Newtonian hypothesis or to construct something like Handel's *Water Music*.[17]

Then Morison quotes Herbert Simon, whom he calls "a cautious student of these matters": "Inso-

[16] Faramelli, *Some Implications of Cybernetics*, p. 219.
[17] *Men, Machines, and Modern Times* (Cambridge: The M.I.T. Press, 1966), p. 77.

WHAT IS MAN MORE THAN A MACHINE?

far as we understand what processes are involved in human creativity—and we are beginning to have a very good understanding of them—none of the processes involved in human creativity appear to lie beyond the reach of computers." Morison believes that the computer "may sometime write a sonata . . . we cannot exclude the possibility that it may feel its own emotion and have a will of its own."[18] In the face of such statements by reputable scholars, I can only suggest again, let's wait and see.

What the Machines Cannot Do— Yet

With that warning in mind, let us now look soberly at some things the machine cannot do, yet. Man is more than the machine, however much he may be analogous to the machine in brain functions. He has capacities which no one even suggests belong to the most sophisticated machines. Charles Darwin, I understand, said, "My mind seems to have become like a kind of machine," but he delayed the publication of *The Origin of Species* some twenty years for fear his wife would be embarrassed by it. Apparently that part of Darwin's mind which decided to protect

[18] *Ibid.*, pp. 94-95.

PLENTY AND TROUBLE

his wife's feelings was not a machine; I do not know of any machine that "feels" for the sensitivity of a woman.

A person speaks about his "self," his "I," which science cannot define and some psychologists deny, but every man experiences "for real." His self may be in turmoil, but he does not deny that something is there. His self is not a thing, not a ghost in the machine, not a possession, not a substance, not a soul that flies away on wings at the time of death. The self is that integrating center of a person, that dynamic sense of "being a person," or "having a life," and that is something that the computer does not have. If you pull the cord from the wall, the computer has no power to pull itself together or even to complain about its condition.

The computer does not worry about wearing out, nor fret about being replaced by a later model. It feels no pain, and has no competitive spirit. Nor has it any sense of motivation when things go wrong, such as Fr. Tyrrell spoke about. "When I am tempted to give up the good fight, the sight of that Strange Man hanging on his cross drives me back to my task again."

Nor does the computer ever bow its head or take off its hat or give any thought to the intangibles. Surely in June, 1940, a computer

WHAT IS MAN MORE THAN A MACHINE?

weighing all the known measurable factors would have advised Churchill, faced with Hitler and the Nazi weapons, to sue for terms, because the computer cannot measure such intangibles as the spirit of the British people and the eloquence of Churchill's rhetoric. Confined as it is to "measurable factors," the computer could make catastrophic decisions.

Harvey Cox is right. "Machines . . . can be astonishingly efficient. But there are some things they cannot do. Among other things, they cannot play, pretend or prevaricate. They cannot frolic or fantasize. These activities are somehow uniquely human." [19] And who will disagree with Joseph Wood Krutch:

Certain enthusiasts . . . predict that since consciousness is merely "alertness," we will someday create a machine so alert that it will become conscious. But that is a mere "someday." Until a computer begins not only to propose and answer questions, but also to debate with another computer whether or not they are both mere machines (and loses its temper in the process), sensible men are justified in doubting that anything resembling a brain has been constructed out of tubes, transistors, and cables.[20]

---

[19] *The Secular City,* p. 28.
[20] "The Brain vs. the Machine," *Saturday Review,* January 18, 1964, p. 19.

PLENTY AND TROUBLE

Symbiosis: The Marriage of Man and Machine

One way to handle the distinctions between men and machines is to say, Render unto the machine what the machine can do best, and unto man what man can do best. In that case, we would give to the machine the work that must be done fast, with exact precision, in great volume; and to men the work that requires subtlety, individual differences, and evaluation of intangibles.

That, however, is too simplistic. It makes false distinctions, and it would deprive us of many projects to which men and machines both contribute, each bringing his or its distinctive capacity. In those cases, we have a marriage, or what biologists call symbiosis. Consider examples in two fields, language and medicine.

In the field of language, computers are coming into increasing use for abstracting and for translating. The enormous explosion of knowledge and writing makes it impossible in many fields for even the most erudite men to keep up with the literature. For instance, the periodical, *Chemical Abstracts,* which gives short digests of articles published in the field of chemistry, consisted of fewer than a thousand pages in 1933; in 1969 it was almost 25,000 pages. Chemists say they

WHAT IS MAN MORE THAN A MACHINE?

need a digest of the digests! [21] Anyhow, the task of writing the digests can now be handed to a language computer. The machine scans the original article; then, based on the frequency of words and the location of words in the paragraph and other criteria, the machine types out from the article a digest that gets at the substance in a concise and useful way. Every preacher knows, of course, that a computer compiled the concordance of the RSV Bible; it certainly looks like it, too—a mechanical, undiscriminating listing of words. It has no human touch, yet that is exactly what the machine needs when it digests or translates. When a computer translates from one language into another, the difficulties arise over the "little" words with multiple meaning or shadings, and the machine must be taught to recognize these nuances of thought. That is a long, tedious process in which a human translator must instruct a given computer. Dr. Norbert Weiner, pre-eminent leader in the field of cybernetics, concludes:

[21] "The plea of the scientist seems to be more for abstracting the abstracts than for improving their literary quality. In our more leisurely past we could take time to explore the meaning of meaning, but since the advent of the knowledge explosion we can no longer know what we know." Hatt, *Cybernetics and the Image of Man*, p. 191. It is sometimes joked that a laboratory technician finds it faster to repeat a research project than to find the account of the original experiment.

PLENTY AND TROUBLE

The best hope of a reasonably satisfactory mechanical translation is to replace a pure mechanism . . . by a mechanicohuman system, involving as critic an expert human translator, to teach it by exercises as a schoolteacher instructs human pupils. Perhaps at some later stage the memory of the machine may have absorbed enough human instruction to dispense with later human participation, except perhaps for a refresher course now and then. In this way the machine would develop linguistic maturity.[22]

Here we have a vivid illustration of the fruitful intermix of human and mechanical elements.

A still more spectacular field for the close interplay of men and machines lies in medicine, where the mix of human and mechanical elements is already able to replace damaged limbs or organs. Suppose a man loses his hand at the wrist, but the major muscles that normally move the hand and fingers remain intact in the stump of the arm.[23] When contracted, these muscles now move no hand or fingers, but they do produce certain electrical effects that can be picked up by electrodes, amplified by transistor circuits, and used to control the motions of the artificial hand through battery-driven motors. That is, the same nervous signal that formerly produced muscular

[22] *God and Golem, Inc.* (Cambridge: The M.I.T. Press, 1964), pp. 79-80.

[23] This illustration is described by Weiner. *Ibid.*, pp. 74-75.

WHAT IS MAN MORE THAN A MACHINE?

contractions now gives off electric impulses that instruct the motor to move the artificial hand. Thus, the use of the artificial feels easy and natural.

The natural hand is an organ of motion, but also an organ of touch. But can an artificial hand ever feel? Yes, it can. Tiny pressure gauges can be put into the artificial fingers, and these gauges can then communicate electric impulses back to the surface of the skin, say at the end of the stump, and the same nerves which once relayed the sense of touch from the natural fingers now are activated by these impulses which come from the identical points of the artificial hand. Thereby we can produce a vicarious sensation of touch to replace the natural sensation.

Thus, the close interworking of human and mechanical elements means we no longer have to insist on the differences of men and machines, nor proudly assert the superiority of men. Instead, we can utilize the machine for all it can contribute to human welfare, and rejoice for its power to extend the power of human mind and muscle. Morison is certainly right when he says that the computer

by its very presence and nature puts the essential questions, defines the central problem . . . what is the nature of man? . . . The computer by its practical

PLENTY AND TROUBLE

consequences can force man, for the first time, to raise up and examine his being . . . to face not only what he is, but what he is doing, why he is doing it, and what he wants to do.[24]

Wonder and Praise for New Creation

We began by saying that man becomes *homo faber in order to be homo sapiens*. He designs machines to magnify his powers. He designs computers to magnify the powers of his mind. Other machines give us wings and fins so we can imitate and excel the birds of the air and the fish of the sea, but the computer extends our power to think and learn and remember in ways no other creature is able to do. When we couple these machines with human skills, and marry men and machines into a mechanicohuman system, we magnify our powers of speech and healing. Surely these traits belong to human persons who are created in *imago dei*. A man is more than a machine, yet the machine helps a man to fulfill his image of God. Render unto the machine, therefore, the works the machine can do best. Render unto man the works he can do best. Render unto the mechanicohuman system the work which neither man nor machine can do alone, but when done together becomes another wonder of God's creation.

[24] Morison, *Men, Machines and Modern Times*, p. 86.

5
Self-made Men

Now is the hinge of history. Everything turns in our time. Within our memory spectacular changes have happened. Speed increased from 100 mph. in the locomotive to 18,000 mph. in the spacecraft. Energy expanded from the gasoline motor to atomic power that blasted a whole island out of the sea. The voice carried by telephone now bounces off a satellite. Travel has reached the mountain peak, the ocean depth, and probed out to Venus and Mars.

Yet still more striking, with vast repercussions, are the new happenings in bio-medicine, where the spectacular truth is this: we have come to the end of evolution by natural selection only. The natural process that determined our destiny for three million years has come to an end during our lifetime. Life had been shaped by accident,

PLENTY AND TROUBLE

disease, plagues and powers of nature, and the unpredictable mutations in the human gene. No longer so! Hereafter human life results from what people do, their choices responsible and irresponsible. We human beings are soon to be self-made men. We will no longer leave it to the snake, or to fate, or to the four horsemen of the apocalypse. We will create our own life as a result of the new technology in the field of bio-medicine.

Let your imagination play over the vast possibilities.[1] "No idea seems too wild to contemplate. What would you like: Education by injection? A catalogue of spare parts? A large, more efficient brain? A cure for old age? Immortality through freezing? Parentless children? Custom-ordered body size and skin color? . . . Name it, and somebody is seriously proposing it."[2] Let us explore only three issues in this vast field, and begin with population control.

---

[1] For a spectacular and quotable instance, see "Ode on a Plastic Stapes," by Chad Walsh, *Saturday Review*, January 22, 1966, p. 33. The poet-patient celebrates the surgeon who replaced his natural stapes with a plastic one, so the creature's skill prevailed "where God's hand shook." Then the poet exclaims, "Praise God who made the man who wrought this wonder."

[2] Albert Rosenfeld, *The Second Genesis: The Coming Control of Life* (Englewood Cliffs, N.J.: Prentice-Hall, 1969), p. 9.

SELF-MADE MEN

The Population Issues

To describe the increase in population, the words "explosion" and "bomb" are not rhetoric.[3] Facts justify the fear that the fuse is already lit for an explosion that will prove deadly to millions of people, if not to mankind. This is a new thing in human history, and the most extravagant language can hardly overstate it. Some people feel there are already too many people alive, and Dr. Barry Commoner, distinguished ecologist from St. Louis, believes that the eight billion population in the year 2000 is the absolute limit for decent living, and he sees no way to avoid at least that number even if we begin the most drastic controls immediately.

The Statistics Are Frightening. To get perspective on the population growth, remember that at the time Jesus lived there were about 250 million people on the earth. It took until 1850 for the population to reach one billion, then only 80 years to reach two billion, only 35 years for the third billion, and by the year 2000, only thirty years hence, there will certainly be at least seven billion people. Each day the sun rises on 191,000 more people than the day before; and every one of these

[3] For instance, Richard M. Fagley, *The Population Explosion and Christian Responsibility* (New York: Oxford University Press, 1960), and Paul R. Ehrlich, *The Population Bomb* (New York: Ballantine Books, 1968).

PLENTY AND TROUBLE

new human souls has a body to be fed, clothed, housed, educated, nursed, healed, governed, provided for. Each year there are 70 million more people, 60 million of whom live in the less developed nations, and these people are not white, Anglo-Saxon, Protestant, capitalist, and democratic!

How did this happen? The answer is simple: technology, medical technology. DDT almost abolished malaria, typhus, yellow fever, and cholera, and the death rate declined all over the world. Countless medical advances improved health and postponed death. It is not that people live longer, but that more people live out their biblical allotment of three score years and ten. In Jesus' time life expectancy was 25 years; today it is almost 70 in much of the world. The major cause of the population growth is the decline in the death rate that results from medical science and public health programs. This is a new thing; it never happened before. The trend is accelerated because more women live to childbearing age, and carry children to full term, and the mortality of childbirth is decreased. Consequently, there are more people, to the point of threat: more people hungry, miserable, illiterate today than there were a generation ago. Such is the bad news that comes from the good thing of medicine! This is perhaps the example par excellence of how the by-products

SELF-MADE MEN

of technology are not foreseen and may be more harmful and harder to cure than the original problem.

Effects of Population Expansion. Despite vigorous efforts to increase food production, it is true that in many lands today the population growth outruns the food increase.[4] This means the ancient biblical promises are not fulfilled. "Lo, everyone who thirsts, come. . . ." but everyone cannot come and drink because water supply runs short. The promised land will not flow with milk and honey because the food supply runs short. People will not inherit the land prepared for their fathers, because there is less land per person than ever before. The United Nations Food and Health Organization estimates that 1.5 billion people suffer from malnutrition—more people than the world's entire population a hundred years ago! In affluent countries the average per capita consumption of animal protein is 44 grams per day; in low-income societies it is 9 grams per day, and the name of that game is starvation. According to the Malthusian predictions, famine will occur. Enormous populations, living hand to mouth, year to year, stand ready victims for any year of poor

[4] The experts disagree whether enough food resources can be garnered from the sea, or agriculture be improved enough, to feed the new billions of people, but most observers expect severe famine within the coming decade, alas!

crop or new disease or natural catastrophe. That is the cruel way to adjust the death rate to the birthrate. A "human glacier moves inexorably in upon us with the threat of a terrible future of misery and servitude." (Everett Dirksen)

Put this population growth into the large perspective of ecology. According to René Dubos,

> the most disturbing problems that will arise from a larger population are probably not amenable to technological solutions. As he populates more and more of the earth, man will have to eliminate all forms of wild life that would compete with him for space and for food; he will increasingly have to flood deserts and fell forests in order to create more farm land, factories, houses, and roads . . . Man will thus destroy all the aspects of the environment . . .[5]

Regimentation will be unavoidable, because nothing is more volatile than a generation of hungry young men. Freedom and privacy will become anti-social luxuries. Thus, the type of human being most likely to prosper will be that person who is willing to accept the regimented way of life in a polluted world from which all wildness and fantasy have disappeared. Such is the gloomy prospect before us.

Some Americans tend to slough off the issue as though it affected the rest of the world, but

[5] "Man's Unchanging Biology and Evolving Psyche," *The Center Diary 17*, p. 40.

not them. Ah, not so. Here in the United States we are a little behind in the race to suicide, but not far. Our population of 206 million will grow, the Census Bureau estimates, to 281 million by the year 2000, based on the 1968 birth rate which was the lowest in our history, or to 321 million based on the birthrate of the 1950's. Our resources are abundant but not infinite. We can likely afford to build schools, houses, roads, and the multiplied services required for this increase, but we also do more damage to the environment than other nations do. Paul Ehrlich estimates that every newborn child in this country is fifty times the threat to the environment of a child born in India.

Supposing the good earth can provide the resources to survive on, where is the enjoyment of living in such congestion? As cities expand further, government services multiply simply to hold things together. College doors close in the faces of the young, and the quality of education slumps. The water supply gets low, further open space is converted into dumping grounds for cars and atomic waste, costs go up, long traffic lines waste more time. It is like living in an elevator, a global squeeze, and there comes a point where change in quantity becomes change in quality. Crowding threatens the quality of life. It is well known that crowding has a very damaging effect on animals. Deer die under the stress of overcrowding, even

PLENTY AND TROUBLE

when food is abundant. Rats in confinement become violent and dirty and sexually deviant, in contrast to their normal patterns of peaceable living, in cleanliness, under established customs of courtship and marriage. To my knowledge there are no reliable data on human beings, but it may not be too wild a guess that people, too, when crowded behave in similar ways.[6]

Controls on Population Growth. This prospect suggests that there must be some controls on unlimited population growth. That very thought runs counter to the traditional human understanding that the decision about having children is a basic right of the family. If we honor that tradition, then population control must be voluntary. Voluntary methods include, of course, those acceptable to the Catholic Church, namely continence and "the natural cycle" method; also the more widely used contraceptives of many kinds; the delay of

---

[6] Studies of Japanese prison camps and German concentration camps reveal aberrant behavior under conditions of crowding, starvation, and stress.

See Dubos, *So Human an Animal,* p. 154. See also Conrad Lorenz's book, *On Aggression,* which maintains that the invasion of animals' "territorial identification" results in a breakdown of their social patterns and even cannibalism of their young.

M. T. Hall, author of *The Hidden Dimension,* believes that each person needs, besides food and shelter, a certain amount of space, a space-bubble, to live in comfortably. To explore this problem he created a new scientific discipline which he calls "proxemics."

SELF-MADE MEN

marriage until the late twenties; the new sterility pills; the resort to abortion, an operation now legal in several states at the woman's request; and surgical sterilization of the male, as practiced in India.

But will voluntary methods prove effective? Many experts in this field believe not. Garrett Hardin believes "voluntarism is insanity. It results in uncontrolled population growth." Voluntary restraint means at least this, that people of conscience and strong sense of social responsibility have small families, while those of no conscience have large families, so those of conscience lose out in the social struggle. Thus, the scrupulous are at the mercy of the unscrupulous. Shall they then become unscrupulous in order to survive? The way of self sacrifice is commended to us on the very highest authority of the Cross, but is it the way for the human species on the crucial question of survival?

Even if we put our confidence in voluntary controls, the question still arises as to whether there is time enough for education before the time of disaster. Or is the problem so urgent that social wisdom requires external controls? Controls might take a variety of forms. Income tax policy could give deductions for two children only, but no deduction, or even a tax assessment, for more than two. Pension rates and old age security payments

PLENTY AND TROUBLE

could be cut for anyone having over two children. Each citizen could be given a license permitting him to have two children only. Then there are scientific methods, such as the drastic method of putting a sterilant in the city water supply to cut down the birthrate.

Christian thinkers have a divided mind about all this. At the Conference on Church and Society, sponsored by the World Council of Churches in Geneva in 1966, Section IV brought in a report with two differing findings. "Responsible parenthood is not just a matter of individual family concern; it must be accepted as an integral part of the social ethic of the day," says Par. 60. Further on, Par. 105 says, "Every couple has a right to make its own responsible decisions on the planning of its own family in accord with its moral and religious convictions." As Prof. Roger Shinn rightly says, anyone familiar with the drafting methods of assemblies seeking consensus can guess that the two paragraphs came from two subcommittees and that the larger group may never have noticed the conflict, and if it had, might have been unable to reconcile them, because the church is ambivalent on this question.[7] Prof. Shinn then points out that every ethical decision involves conflicts of value.

[7] An unpublished address, "Population and the Dignity of Man," delivered before AAAS meeting, Boston, December 28, 1969, p. 15.

SELF-MADE MEN

A desperate world may use coercion to limit population, as single societies have sometimes done in the past. But the dilemma is a bitter one. . . . Human dignity demands limitation of population. But some methods of limitation destroy dignity. Infanticide, for example, is as bad as any problem it is designed to solve. It may remind us of the army officer who explained that he had to destroy a village in Vietnam in order to save it. . . . Life permits no total freedom or total coercion. Society, if it is to survive, will probably learn to limit population by means of persuasion and pressure that fall somewhere between uninhibited freedom and overt coercion. Many methods of persuasion and pressure are possible. Prestige systems, economic pressures, taxation, housing policies, and skillfully contrived propaganda are a few of the devices by which societies are likely to move increasingly as they see the necessity of limiting reproduction. It is not wrong for society to use such pressures. Society itself is under immense pressure; there is no reason why the families within it should evade the pressures.[8]

[8] *Ibid.,* pp. 17-18. See Garrett Hardin, "The Tragedy of the Commons," *Science,* Vol. 162, Dec. 13, 1968, for further argument that controls over population growth are similar to the necessary controls over bank-robbing, parking, and other public issues. Hardin recommends "mutual coercion mutually agreed upon." This attitude was endorsed by Louis Hellman, deputy assistant secretary of Health, Education and Welfare, who said to the American Medical Association, meeting in Boston on December 2, 1970, "There is no evidence that voluntary family planning will have any significant effect on the population growth of the United States. We must go one step beyond family planning, while maintaining the absolutely essential voluntary aspects of the effort."

PLENTY AND TROUBLE

Dr. Barry Commoner believes that the only effective way to limit population growth is to raise the standard of living. People on the rise, people who see a good future ahead of them, people who hope for comfort and decency for their children—such people do, as a matter of fact, restrain their family growth. "History has shown," he argues, "that people stop having large numbers of children only when they gain a measure of security." Dr. Commoner suggests, therefore, that American people voluntarily reduce their own standard of living to share their wealth with the underdeveloped peoples. Ah, there's the rub. Some Americans would be willing, but can such idealism become effective in public policy soon enough to prevent the population explosion that threatens to blast mankind off the earth?

By one means or another, or by a combination of them all, the world community must slam on the brakes to its own growth and come to approximately a stable population. That means ZPG, zero population growth.

The Judeo-Christian ethic of the family was formulated in a time when there was a shortage of people. "Be fruitful and multiply" was a policy that brought economic and social benefits. But now further population growth will starve millions and endanger the quality of life for everyone. This

SELF-MADE MEN

makes a wholly new problem for the human race, because *the rate of human conceptions* has been rather constant, it is believed. It is the dramatic drop in infant mortality and adult death rates that makes the population explosion. For remedy, society is forbidden by its humaneness to revert to brutal *death* rates. The only alternative is to create new controls over *birth*. This requires a new ethic, which is, of course, a religious issue, and it illustrates again the severe tension created for religious people by the new technology.

Transplantation of Human Organs

We now turn to the second major issue in the field of bio-medicine, and nowhere is modern technology more spectacular than in the transplantation of human organs. Some three thousand patients have received kidney transplants, and about two hundred, heart transplants. There have been a few liver and lung and bone marrow transplantations. This represents a striking new direction in medical practice. "Medical *techne* in the Hippocratic sense . . . originally confined itself largely to prophylaxis and the techniques of protecting the integrity of the human organism, to maintain or restore its stability." But today the "new physician" is also "venturing out into the

PLENTY AND TROUBLE

manipulating intervention and interference. . . ."[9] This new, expanded concept of what medicine is called to do, vividly illustrated in the transplantation of organs, has had a resounding impact on people's thinking. They now feel that almost anything is possible. They talk about human plumbing, and spare parts, as though the human body were a machine that can be readily repaired. Transplantation has had the further result that it sharpens up the ethical issues for widespread public inspection. Dr. Henry K. Beecher, of the Harvard Medical School, who has reflected on these questions for many years and written widely for the medical profession, says:

Some generalizations can be supplied as useful guides in the transplantation field. . . . It may be helpful to summarize them. . . . There must be a reasonable chance of clinical success. An acceptable therapeutic goal must be present. . . . The risks and uncertainties must be presented to the families of the donor and the recipient as well as to the two principals. The protocol for each transplantation must be devised so as to gain and preserve the maximum information. There must be a probing evaluation of the results by independent observers. Careful, accurate, conservative information is to be disseminated through legitimate channels, both

[9] Paul R. Ehrlich, "The Biological Revolution," *The Center Magazine* II, no. 6 (November, 1969): 34.

SELF-MADE MEN

medical and lay, in order that cruel hopes will not needlessly be raised.[10]

There may have been some violations of those guidelines, especially of the last one, but I believe Dr. Beecher is justified in saying these guidelines have been generally observed by responsible investigators. The ethical problems lie elsewhere, and medical men are even more aware of these than the public is.

Informed Consent. It is generally agreed that both patient and donor must consent to the transplantation, and say yes out of full knowledge of what is involved and what the prospects are for success. Yet such informed consent is very difficult to obtain. To begin with, the patient is often in such condition that he is unable to make such a serious decision. Furthermore, even in sound mind and good health, the patient cannot understand "everything" involved. Also, he tends to overestimate his chances, and to block out bad news. Out of sheer desperation he sees more hope than there is. Beyond that, the patient is tempted to agree to the transplant even at great risk because he wants to please the physician who wants to do

[10] "Scarce Resources and Medical Advancement," *Daedalus*, Spring, 1969, p. 296. Reprinted by permission from *Daedalus*, Journal of the American Academy of Arts and Sciences, Boston, Massachusetts, Volume 98, Number 2.

it, and because he fears incurring the doctor's disfavor if he refuses. In any case, even informed consent does not relieve the doctor of his responsibility, nor the family of its, and the family must oftentimes make the decision when the patient is a child, or elderly, or incapable of making the decision himself, as in the case of shock from serious injury. The ethical question then becomes: What measure of informed consent is required, and who is responsible for giving consent?

Injury to the Donor. Another ethical question arises concerning the bodily damage done to the donor. The damage may seem slight, as in the case of kidney donors. The increased mortality among donors who retain one healthy kidney is less than two per thousand over the next five years. "Translated into survival for a healthy 35-year-old male donor (of a kidney) it means an expectation of living out the next five years of 99.1 percent or more as compared with a normal expectation of 99.3 percent. This figure is comparable to the increased risk in commuter driving for a total of only 16 miles per working day."[11] (Prof. Augenstein tells of a legal contract in which one

---

[11] John P. Merrill, quoted in John Holden, "Some Ethical Considerations in the Transplantation of Organs," *Existential Psychiatry,* Summer, 1966, pp. 173-87. It is harder to estimate the long-term risks; "they are probably small, but not negligible."

SELF-MADE MEN

man, with complete kidney failure, offered to pay $10,000 for one good kidney. "However, the contract was not for an outright sale, but rather a lease arrangement," and the donor was to retain the right, in case of the failure of his remaining kidney, to retrieve his leased kidney.)[12]

Injury to the donor is not primarily to his body, but to his emotions. As a prospective donor he is not free. What brother can refuse to give a kidney to an ailing brother or sister? Even a close friend cannot make a free decision if asked by a sick friend. One man who investigated this matter found considerable disruption of mental and emotional health among potential donors and their families. "I know of a patient's brother who declined to donate his kidney—with resultant severe emotional trauma; I know of another family torn apart by a mother giving a kidney to her child against the wishes of the husband and father.[13] Clearly it is necessary to protect children against such trauma, and public morals makes it unadvisable to permit prisoners, or medical students, or other categories of captive or "related" persons to volunteer their organs for transplant.

Further problems arise. For instance, to take an

---

[12] Leroy Augenstein, *Come, Let Us Play God* (New York: Harper & Row, 1969), pp. 41-42.

[13] J. Russell Elkinton quoted in Holden, "Some Ethical Considerations," p. 17.

organ from a healthy person appears to contradict the doctor's Hippocratic oath that he will do no injury to his patient. Clearly there is injury, and some risk, when any organ or bone marrow or such is removed.

Again, there is the question of the donor's right to sacrifice himself. At first glance such donation appears noble. In war we decorate the soldier who risks his own safety and gives his life to protect his buddies; if a man may give his life may he not give an organ? In peacetime we applaud the fireman who returns to a burning building to rescue a child; and we honored Capt. Oates who walked out into the Antarctic blizzard so his companion could survive on food rations too meager for two men. And we remember Jesus Christ our Lord who laid down his life for the brethren. Does anyone dare suggest it is wrong to sacrifice oneself? "Greater love has no man than this, that he lay down his life for his friend." (John 15:13)

Take a dramatic case: a mother offers to donate her heart to her own child. Now the heart is a single organ, you have only one, and donation of the heart means sacrifice of life. Should this be allowed? The most insightful comment I know is a book with a chapter entitled, "The Protection of the Non-Tipper," which takes a clue from the Faculty Club at Berkeley which forbids tipping the waitress. That rule aims not so much to pro-

SELF-MADE MEN

tect the dignity of the waitress as to make things comfortable for the member who is either unable or unwilling to tip. If others tipped he would be embarrassed and pressured into doing what he disapproves. The author suggests other illustrations of this same principle, such as uniforms in schools and limitations on election campagin expenses, to protect those who cannot afford elegant clothes or extravagant campaigns.

I submit that the protection of the non-tipper lies at the root of the unanimous rejection of the donation of an unpaired organ for transplantation. . . . If a person, alive and healthy, could be approached to surrender a vital organ to a child, parent, spouse or friend who, but for this gift, must die, it would put him under fearful pressure. He would rarely bring himself to comply, but refusal would leave him wretched. . . . The tension would be unbearable. That is the true reason we hold any such donation illegal.[14]

Alternatives to Transplants from Living Human Donors. Medical people could avoid some of these problems if other donors could be used. Animals, for instance. One writer asserts that "most medical scientists are agreed that the use of animal or nonhuman sources for transplantation is preferable to the use of human sources." For my part I see no

---

[14] David Daube, "Limitations on Self-Sacrifice in Jewish Law and Tradition," *Theology* LXXII, no. 590 (August, 1969): 303.

PLENTY AND TROUBLE

particular ethical issues here beyond the restraints that apply to all use of animal life for scientific purposes. Also, human cadavers might be used, if ways can be found to "store" human organs or whole bodies for later use.

In any case, the success of transplantation depends largely upon solving the body's rejection of foreign material. The body produces white corpuscles which fight disease and reject foreign intrusions. When the intrusion is a transplanted organ this natural defense becomes an obstacle to success. Modern medicine uses drugs, tissue-typing and radiation (called immuno-depressant therapy) to cut down the natural rejection, but that treatment also reduces the body's resistance to infection, hence endangers health. The ethical question becomes: Should we force the body to contradict its own genius?[15]

Another possibility lies in the deliberate growth of new organs; that is, the cultivation of a new heart, a new liver, to replace the damaged one. That possibility is still doubtful, though it is being seriously discussed. In any case, it is far off in the future, although Augenstein makes the bold prediction "that we shall be able to remake at least hearts and livers within this century."[16] Once

[15] Kenneth Vaux, "The Heart Transplant: Ethical Dimensions," *Christian Century*, March 20, 1968, p. 355.

[16] Augenstein, *Come, Let Us Play God*, p. 45; here he explains the biology of this prospect.

SELF-MADE MEN

again we come to the frontier question: Does all this mean that a man is a bundle of parts to be replaced or regrown as they wear out? Is there no limit to the repairs that are possible? Do we need not admit death as "the *last* enemy?"

Who Gets the Few Available Organs? When there are more patients in need of transplants than there are donors available, who get chosen for the gift? The wealthy who can pay the enormous cost, up to $80,000 for a new kidney for instance? Or the most talented and those most likely to make significant contribution to society? Or those with large families, who presumably are most needed at home? What are the guidelines to help a doctor make such painful decisions?

First of all, there are medical practices. Prof. M. F. A. Woodruff has listed some necessary conditions under which it is permissible for a living volunteer to donate a kidney: the transplant must be a last-resort procedure; the donor must be healthy and not likely to suffer from the donation; he must be a free-choice volunteer; there must be informed consent by all parties; and there must be a reasonable chance of success.[17] Yet after these guidelines are observed, the puzzling

[17] "Ethical Problems in Organ Transplantation," *British Medical Journal*, 1964, I: 1458; the exact language of these guidelines is quoted in Holden, "Some Ethical Considerations," p. 181.

PLENTY AND TROUBLE

question remains: Who among the needy patients is to be chosen for treatment? The ancient and honorable lifeboat-in-the-storm principle might be used: first come, first served. There is biblical precedent for casting lots. In such non-choice there will be factors of luck and fate, yet perhaps a lottery will be the closest approximation we humans can make to the Divine Providence that loves all people equally.

Definition of Death. This brings us to the very tricky question of defining death. The medical situation in transplantation requires that the organ be "fresh." This means, in the case of the kidney, within three hours after removal from donor or cadaver. In the case of the heart it means almost immediately after death. Therefore, the surgeon must know the exact moment of death, and be prepared to act promptly. Clearly the questions arise: When does death occur, medically? And may the moment of death be controlled for the benefit of the patient scheduled to receive the transplant?

In former times death was determined by holding a cold dry mirror to the lips, and if any detectable breath condensed, there was still life, and hope. Today that method is not adequate because the heart can be invigorated by use of drugs, massage, and electrical pacemakers, and the lungs kept breathing with respirators. A few years ago a woman was "clinically dead" for forty minutes before she responded to adrenalin and chest massage, and returned to normal life.[18] Was she really "resurrected from the dead"? If so, that raises severe questions about some gospel miracles!

Another indicator of death is the electroencephalograph which indicates the electrical activity of the brain. The living brain gives off electrical impulses which can be recorded and their pattern ("brain waves") may be studied for indications of health or pathology. If the graph is flat it means the brain is dead. But brain death does not always occur simultaneously with the stopping of heart and lung action. Therefore, conservative medical practice today requires that all indices of death must concur before death can be pronounced. Five tests are standard. The heart and the lungs cease to function, the eye pupils dilate, the body gives no reflex response (such as the knee-jerk), and the brain registers a flat line on the electroencephalograph. When all five functions cease, the person is dead.

Clearly it is possible in many dying patients to "adjust" the moment of death. By mechanical and medical methods the patient can be kept "breath-

[18] AP story from Santa Barbara, Calif., *New York Times*, August 10, 1964.

PLENTY AND TROUBLE

ing" and his heart kept "beating" for a considerable time in some cases. By turning off the machine the patient then dies within a few moments.[19] The ethical question becomes: Is it permissible to "set the time" for death, something as we set the time for birth when a Caesarian operation is scheduled for next Monday at 8:00 A.M.? How long, if at all, is it morally permissible to maintain life in a patient who has suffered irrevocable brain damage, for instance?

The question of death and its control arises sharply whenever an aged person comes to a terminal illness. Augenstein tells of his own grandmother who, at age 89, went into a coma. The doctor stated that by heroic methods ("heroic" means extreme, all-out methods to sustain life) he could keep her alive a few weeks or months, whereas if he "made her comfortable" she would live, at most, a few days. Since she had deteriorated badly both in body and in mind the last few years, and was obviously unhappy and uncomfortable, her son told the doctor to "make her comfortable," and the author, Augenstein, concurred with his father in that decision.[20] That

---

[19] For a popular account of an actual incident in which the patient was maintained for over twenty-four hours to keep his kidney alive for a transplant, see Leonard Stevens, "When Is Death?" *Reader's Digest*, June, 1969, pp. 88-89.

[20] Augenstein, *Come, Let us Play God*, p. 50. The distinguished Dr. Charles W. Mayo draws on vast experience

SELF-MADE MEN

decision will confront many, many people who love their parents intensely. The new technology requires them to decide whether to keep a person alive or to let him die. The presumption is that the medical team will seek to keep the patient alive; that is its business. The burden of proof will fall on those who allow death to happen if they have available any reasonable means to prevent it. But today "reasonable means" include drugs and machines which help people to "keep alive" for weeks and months after life has ceased to be enjoyable or meaningful. Life may in fact become painful under these circumstances; the patient may become totally dependent and unable to care for himself; he may be out of touch with reality and unable to communicate by word or sign what he is thinking or feeling, if indeed he is thinking or feeling anything. In that case, the decision to extend life is really a decision not for life in any meaningful sense but a decision for mere existence. Does life consist of mere existence,

---

to express his opinion: "In recent years, medicine has advanced greatly in techniques for maintaining some kind of life in the dying for years at a time. I think we have gone too far in this direction. We should keep people going only so long as it is possible to restore them to a sane, conscious degree of health. It is a tragedy to maintain life in an unconscious vegetable, but it is happening in every hospital in the land; it drains the families economically and emotionally, and serves no purpose that I can respect." *The Story of My Family and My Career,* Reader's Digest condensation, p. 374.

PLENTY AND TROUBLE

or has life, in the meaningful sense, ceased somewhere back along the line? And who makes this decision, the doctor or the family? We are not talking about euthanasia, which is purposeful death to relieve suffering. That is an act of killing, however merciful in intent, and I find it hard to make a Christian case for that, though I do not deny it categorically. Here, however, we are talking about the decision forced upon us by the availability of new technology, the decision whether to use "heroic" methods. It is necessary to distinguish between euthanasia and allowing the process of death to proceed. Care for the dying does not require that we interfere with the process of dying. Physical life is not the highest value. The martyrs, and Jesus hanging on his cross, allow us never to forget that.

The Cost of Transplantation. One final question, briefly, concerns the cost of transplantations; the economic cost, but more than that, the cost of talent and medical resources. For instance, in New York recently a patient was given a transplant of heart and both lungs, and a team of fourteen surgeons did the work. I wonder if a society that is short of medical people (twenty percent new doctors each year come from foreign countries, part of the "brain drain" from other lands) is justified in investing that much talent in one spectacular procedure. I wonder, when millions of our citizens need better medical services.

Transplantation of human organs may have the further effect that it encourages people to think of themselves as repairable machines. (I said to my wife one day, "My hair recedes, my eyes go blurred, my ears go dull, my teeth decay, my jowls sag, everything goes wrong clear down to my athlete's foot. But I have handsome eyebrows. Why don't I keep them and get everything else replaced?" So feels the technological man!) Whether transplantation heals human disease and thus further develops the healing ministry of the Christ, or tends to cause men to think of themselves as the subjects of technological experimentation—that question remains unsolved.

Genetic Engineering and Other Forms of Human Controls

In the vast field of bio-medicine, perhaps the most fascinating possibilities, and certainly the most frightening, arise from the prospect that we will be able to create human beings to our own specifications. By means of medical technology we will create self-designed, self-made, self-controlled human beings! God made man in his own image. Someday soon man will make man in the image of whatever man he wants!

PLENTY AND TROUBLE

The spectacular discovery a few years ago of DNA, the basic genetic material, means that nature is giving up the final secrets of heredity. Scientists have found the coiled structure, the so-called double helix, of the DNA molecule, thus learning to read the genetic code. Once able to decipher it and read it fluently, they will soon be able to write in that language, hence to give instructions in the DNA code. Reading, then writing, then control—that is the progression, and when that comes, human powers will be godlike. Three young Harvard biologists have recently extracted for the first time a pure gene from a living organism, and they immediately warned that this spectacular advance carries the danger that genetic engineering may be misused to mold the physical and even the behavioral characteristics of future generations of mankind. More recently a team of scientists has achieved the complete laboratory synthesis of a gene.

Eugenic Controls. The first possibilities are found in eugenic control. *Homo sapiens* can be modified somewhat as field corn and beef cattle have been improved by scientific breeding. Each of us carries in his body some defective genetic stuff, and now, by chemical and electronic influences, it is possible to change this material so as to "breed out" the damaging traits and to prevent the disease, deformity, and other defects that are carried in those genes. This is called negative, or preventive, eugenics.

Larger and more troublesome possibilities lie in positive, programmed eugenics, which has been popularized by Huxley's *Brave New World* and espoused by Mueller and Lederberg in this country and many other geneticists the world over. Positive eugenics means planned procreation by carefully selected parents, under the guidance of geneticists. One method is the use of frozen sperm and ova. Already, human sperm are frozen and thawed out for the artificial insemination of women, and hundreds of healthy children playing on the streets are proof that this method works beautifully. As soon as the female ova can be frozen successfully, then either a genetic mother or a genetic father, or both, could be selected for the prospective child. Moreover, the new technology provides chemical and biological influence upon the embryo to change its basic nature, for many responsible biologists believe that such traits as aggression and rage may be subject to control, and the social virtues of cooperative spirit, and specified talents such as music and math, might be "bred into" the embryo. All this infers that some ideal genotype exists in the mind and planning of the geneticist. Hopefully, he will be concerned about "love, joy, peace. . . ." and the other

PLENTY AND TROUBLE

attributes which Paul credited to the Holy Spirit.

Thus eugenic engineering means that defects are erased and talents enriched in the human gene by means of planning and medical manipulation. "No one would argue that man couldn't stand some improvement," as Alfred Rosenfeld put it, "but having the actual power to do so presents some sticky choices. Who is it that we will appoint to play God for us? Which scientist, which statesman, artist, judge, poet, theologian, philosopher, educator—of which nation, race, or creed—will you trust to write the specifications, to decide which characteristics are desirable and which not?" [21] I don't trust anyone, for sure. I don't even trust myself!

This eugenic control can be performed inside the body or outside! In Italy a Dr. Petrucci worked carefully over a "nondescript, blubberlike blob" which was, in fact, a tiny human embryo. He began with a female ovum, then admitted male sperm one of which fertilized the egg, and under his nursing that egg grew into an embryo and grew for fifty-nine days. Conceivably the hu-

[21] *Life*, October 1, 1965, p. 100. Theodosius Dobzhansky, the distinguished biologist, asks the same question: "Are we to have, in place of Plato's philosopher-king, a geneticist-king? And who will be the president of the National Sperm Bank and of the National DNA Bank? What checks and balances are to be imposed on the genetic legislative and the genetic executive powers? Who will guard the guardians?" "Changing Man," *Science*, January 27, 1967, p. 413.

SELF-MADE MEN

man embryo might grow to full nine months *in vitro,* that is, in glass, in a laboratory.

An American biologist received by air shipment from Germany a female rabbit, but inside the rabbit were one hundred incipient prize sheep, all of them embryos only a few days old, growing as if still inside their natural mothers. The embryos had been removed, implanted in the rabbit for transport, and then each was implanted in a ewe sheep where it gestated and later was born. If in sheep, why not in humans? It will be. Biomedical men expect to transfer an embryo from one woman to another.

When you put together all these eugenic techniques, you may speculate with Dr. E. S. E. Hafez that within fifty years it could be possible for a housewife to

walk into a new kind of commissary, look down a row of packets not unlike flower-seed packages, and pick her baby by label. Each packet would contain a frozen one-day-old embryo, and the label would tell the shopper whatever details could be predicted about the probable characteristics of the child. It would also offer assurance of the freedom from genetic defects. . . . After making her selection, the lady would take the packet to her doctor and have her newly-purchased prefab embryo implanted in herself, where it would grow for nine months, like any baby of her own.[22]

[22] Rosenfeld, *The Second Genesis,* p. 125.

PLENTY AND TROUBLE

Consider the implications. Both men and women could ignore the heredity factors in the choice of a mate! (The beautiful but dumb could become popular again!) A child might be conceived of two parents who had never known each other, either or both of whom might be long since dead, and be nourished in the womb of still a third person! Then the riddle of "Whose wife will she be?" becomes an actual puzzle of "Whose child is this?" For the child grown *in vitro,* in that no man's-no woman's land of the laboratory, the existential torment of "Who am I?" becomes no idle question. We are talking about a time when virgin birth is a common occurrence, and women give birth to other women's children. It is the time when romance and genetics are separated, and some favored men may father thousands of babies. "Can the traditional family—already a shaky institution—survive all this?" (Rosenfeld) Think of the repercussions of this new technology upon the family. It abolishes the traditional fact that a human being has two parents, one male and the other female, and that his procreation depended upon their physical uniting, and he grew in his mother's womb until his birth. All that is threatened. What God joined together man is putting asunder. The gospel and the wedding ceremony may both require revision.

SELF-MADE MEN

New Parts and Better Functioning. After the prenatal improvements are made, human beings still fall victim to disease, accident, and death, but the new technology promises many kinds of new relief from such fate. Consider one: new body parts. Transplants, as we discussed, are already here for kidneys and heart, and some for liver and lungs; the stomach comes next, perhaps even the brain in the far distant future. Artificial organs may have some promise. The heart, for instance, is a pump and it might be driven by a tiny atomic-powered battery, something as the pacemaker already regulates the natural heart. Already we use plastic valves and metal sockets. Likewise, the human body can be taught perhaps to tolerate animal organ transplantations. Already ape kidneys, pig livers, and a chimpanzee heart have kept human patients alive for short periods of time. Thus, it may not be too fanciful, with only a little tongue in cheek, to foresee the day when a man will walk around with a plastic cornea, a few metal bones, Dacron arteries, a borrowed kidney, a pig's liver, an artificial heart, and a computerized electronic device to substitute for weak muscles—a kind of Rube Goldberg contraption, a walking assortment of parts held together by some plumbing and willpower and a little bit o' luck.

While technology cannot transplant the brain,

PLENTY AND TROUBLE

it certainly can improve the intelligence. The brain itself may possibly be enlarged by genetic change or prenatal treatment. Short of that, we might do more with the gray matter—really a reddish matter—we already have. Drug therapy and electric stimulation offer great hope that we can improve the brain function, such as remembering more of what we once knew. It has been discovered that in lower animals the training in certain skills makes changes in the structure of molecules in the brain cells. In one experiment, flatworms that had learned to negotiate a maze were ground up and fed to other flatworms, which thereby were able to learn to negotiate the maze much faster than those "on a less educated diet." Therefore, scientists speculate that subject matter might be taught by injecting into the brain an artificial virus containing the nucleic acids of the desired molecular change. Thus, a virus might give us knowledge of German rather than the German measles, and Asian history instead of Asian flu.

Degeneration of the Genetic Pool. One serious threat is the prospect that all this medical work serves ultimately to preserve the unfit, to pollute the genetic pool of the human species. That is, a great many persons carry in their genes disabilities ranging from myopia and diabetes to imbecility. Medical technology cannot yet pre-

SELF-MADE MEN

vent such deformities and defects, but it does preserve such people alive, in fairly good health, and thus makes it possible for them to produce children and pass on to the next generation the mutant genes that produce these defects. In former times, such people would have died by natural selection. Now they are preserved, thus perpetuating and spreading the susceptibility to disease and disability. The arts of healing that are a blessing to the individual may be, at the same time, disastrous to the human species. A few dollars spent on DDT to rid the village of malaria can result in such population growth that starvation results in the next generation.

Professor Hermann J. Muller was so pessimistic about this trend that his warnings were called a "genetic apocalypse." Here we see a vivid instance of the conflict between Christian compassion and social good. Whenever we care humanely for the increasing numbers and an increasing proportion of the human race who are defective, disabled, or deformed in some serious way, we are adding to the burden of a species which is exhausting its space and food and natural resources. It adds up to a gloomy prospect, and results directly from the new technology in medicine. The genetic technology raises the sharp moral question: Is it better to keep defective human beings alive at the risk of severe racial damage, or to sacrifice a few for

PLENTY AND TROUBLE

the possible benefit of future generations? Who are to make such decisions, and by what criteria?

A New Ethic and an Old Faith

In this chapter we have dealt with issues of population, organ transplantation, and genetic engineering as three illustrations of the impact the new bio-medical technology makes upon the practical life and the moral confusion of modern people.

These tremendous problems have developed because technology accomplished something intrinsically good. They illustrate one of the greatest ethical questions of our time. Everyone agrees that good ends do not justify bad means. The tragedy arises when good means result in catastrophic ends. This is a dilemma that requires a whole new type of ethical thinking.[23]

As Charlie Brown put it, "The theological implications alone are staggering." Hence the moral questions. In the Christian community we need a new ethics of genetic duty, just as we need an ethics of abundance, an ethics of care for the earth, an ethics of population control. The new technology has created this need for new moral

[23] Aldous Huxley, "Achieving a Perspective on the Technological Order—a Commentary," in Carl F. Stover (ed.), *The Technological Order* (Detroit: Wayne State University Press, 1963), p. 256.

SELF-MADE MEN

insight. Considering our careless disregard for the future, we human beings have little reason to expect anything great of ourselves. We spoil the earth, we slaughter our finest young men in battle, we let millions starve. In view of this culpable disregard for the coming generations, are we likely to become sensible and charitable toward the genetic improvement of our race? Centuries ago Jeremiah asked the same question:

> Can the Ethiopian change his skin
>   or the leopard his spots?
> Then also you can do good
>   who are accustomed to do evil. (13:23)

Jeremiah implied, with more than a tinge of sarcasm, of course not! The habits of evil are so ingrained that men will not do good. But if it turns out in our time that human beings can change not simply their skin color but almost anything else in their physical, emotional, and intellectual life, then it might follow that they become able and willing also to do good. I do not want even to suggest that we can genetically manipulate ourselves into decency. I repudiate such a statement, which was made recently in a panel of scientists and theologians at an AAAS Conference in Boston in 1969. One medical man said that "when the biological problems are solved, the theological issues will dissolve away." We laughed him out of court.

PLENTY AND TROUBLE

Another biologist seriously proposed that the only escape from universal destruction was the development of a strain of human beings who could establish a living colony on the moon, where the rigors and the purity of the environment would enable men to be completely decent and humane. We laughed at him, too. Wisdom comes from the microbiologist, René Dubos:

The myth has grown that because man has an infinite capacity to adapt to changing environments, we can endlessly and safely transform his life and indeed himself by technology. In reality, there are biological and psychological limits to man's adaptability, and these should determine the frontiers of technological change.[24]

The scientists speculate about the character of the ideal man, the self-made man who might be developed through the new medical technology. Prof. Glass suggests as genetic goals "sound health, high intelligence, general adaptability, integrity of character, and nobility of spirit." H. J. Muller wants to achieve "genuine warmth of fellow feeling and a cooperative disposition, a depth and breadth of intellectual capacity, moral courage and integrity, an appreciation of nature and of art. . . ." Other people believe the ideal

[24] *So Human an Animal*, p. 164.

SELF-MADE MEN

man is not yet visible, but has been seen in the person of Jesus Christ. There seems to be nothing in the plans of geneticists to surpass the moral grandeur of Christ.

Can such a new man be achieved? Suppose science develops in men a larger brain and better memory and keener powers of understanding, will men then be wiser and more humane? Or is it still safe to say that though a man travels to the moon he is not necessarily any closer to the kingdom of God? And though he restructures his genes into some ideal genotype, he still is *imago dei,* still a sinner, still a forgiven sinner, still a person whom God seeks and finds, and holds responsible.

6

Some Pro's and Con's

Our conversation thus far has delved into some history, some future-guessing, and some description of technology. Then it ranged over three areas where the impact of technology is felt most vividly. Now we undertake a tentative evaluation —a listing of some characteristics that make technology a cause for laughter and dancing, and others that bring tears and cries of pain. First the undeniable benefits.

1. *New Abundance of Goods and Services.* The new technology produces a cornucopia of goods: more food, more medicine, better food, better medicine; faster and safer travel; more information, more easily accessible; better teaching aids and better entertainment—a vast range of goods and services which mean health and enjoyment and comfort for multitudes of people.

SOME PRO'S AND CON'S

The Christian man will recognize this technological advance as a gift from God. He will rejoice for it because it remedies some of the hunger and disease and loneliness and misery and ailments human flesh inherits. Christianity has been called the most materialistic of all religions because it knows the world of material things as the place where God works, where the Messiah deals with his people, where Jesus talks about feeding and healing, where bread and wine are means of grace. Therefore anything that improves the material welfare of God's people on God's earth deserves approval. Technology qualifies.

2. *To Serve the Neighbor's Need.* Technology makes it possible for Christian people to obey the ancient command to love their neighbors as themselves. Jews and Christians have long been compassionate in spirit, but until recently they never had the power to make their compassion effective. They spent most of their energy plowing the land and sweating in factories just to provide enough things to keep themselves decently alive and comfortable. Now with technology they have new ways to do all things necessary for themselves and to produce enough to share with all who need. To feed the hungry they now have food processing, hybrid grains, chemical fertilizers, refrigeration, and packaged foods. To heal the

sick they have new medicines, the oxygen tent, and the artificial lung. Now, for the first time in human history, Christians are able to do what they have long been commanded: to relieve the suffering of their fellowmen. "Greater things than I do, you shall do," said the Master. Now, because of technological advances, it becomes literally possible to teach, to heal, and to feed on a vaster scale and with greater skill than Jesus could command. The real world has become potentially a humane world.

3. *Freedom and New Options.* Another benefit of technology is this: it relieves some unnecessary suffering and sets men free from incessant toil. Here in America people readily forget, if indeed they ever knew in any deep, personal way, that multitudes of people spend their days in grinding poverty, unceasing toil, body-wrecking disease, and spend their nights crying themselves to sleep. Now the new technology brings relief, and new life opens up. Huge machinery releases men from backbreaking toil. Delicate machinery saves their eyes and nerves from tedious work. New occupations increase the job choice, giving freedom to choose new friends and a new way of life. Television and telephone bring both privacy and ready communication. Books, recordings, and movies give new choices for entertainment, and

SOME PRO'S AND CON'S

once men are set free from the daily struggle to stay alive, they can turn to other matters more distinctly human: that is, to friendship, study, arts, play, travel. In the more rigid societies where men are still bound to the class structure, the new technology allows people to break out of that confinement. "The technological society puts a premium upon achievement, on what a man does, rather than upon the ascribed place into which he is born. In this way the crust of a caste society is broken." (Paul Deats, Jr.)

When he addressed the World Council of Churches Conference on Church and Society in 1966, Prof. Mesthene praised technology because, "We have the power to create new possibilities and the will to do so. By creating new possibilities, we give ourselves more choices. With more choices, we have more opportunities. With more opportunities, we can have more freedom, and with more freedom we can be more human. That, I think, is what is new about our age."

4. *The Spirit of Achievement.* Another distinctive benefit coming from technology is the spirit it generates in people, an intangible but notable thing. It begins with amazement. Do you remember—indeed, can you ever forget—those hours one night in July, 1969, when you sat before your television transfixed and watched two men set

PLENTY AND TROUBLE

foot on the moon and then dance and bounce across that barren waste? I remember the time my son came home from seeing a medical film showing the inside of a human lung, and the breathing, pulsing tissue of the human body. He was ecstatic about that technical achievement of color movies inside a person. Are we not amazed at what man has wrought?

Myron T. Bloy has written about the buoyant dynamism of the technological world and its spirit of play. The technological spirit commonly says, "We can do anything we want to do." If any problem can be defined exactly, give us a team of research scientists and a federal grant, and we can solve it! Can we get to Mars? Can we control human personality? transplant the human brain and make artificial limbs that feel? Can we control the weather, and make the inner city habitable? Of course we can. We will not be surprised to do these things; we will be surprised if we cannot! This is the technological spirit, the achieving, dynamic, buoyant, we-can-do spirit. Theologically, this means a profound sense that the world is meant for good, and it responds to human reason and work; it provides abundantly for our needs and it is here for our use. Use it! It is good! Hard technical work will produce the good life for all! This dynamic attitude means that men hold a deep, unconscious trust in creation. They really

SOME PRO'S AND CON'S

believe the earth will sustain the good and provide for the future. This secular hope behind all specific hopes arises out of the technological spirit.

That spirit includes the spirit of play. Ask the man in research what he is doing and he is likely to reply, "Oh, just playing around." He likes to tinker with his machine and to watch it work. At his computer console with its tapes and scopes and flashing lights, he feels like a conductor with a baton directing a mighty orchestra. There is an invigorating spirit in the machine; you can play with it and make it do all its tricks. A computer programmer said to me recently, "It rejuvenates me. It's a kind of Fountain of Youth." There was a man paying homage to life itself, making an unthinking celebration of existence!

When we add up the benefits which accrue from technological achievement it seems meet, right, and our bounden duty that we give thanks for the secular vision. Technology fills the future with promise of good things. Technology has become the cornucopia pouring out a wealth of goods and services. Technology has liberated masses of people from superstition and ignorance and drudgery. Technology has aroused in all men a dream of a good life on the good earth, a life that is healthy and comfortable and abundant with goods. Technology has become the system men believe in for an explanation of how things work and how problems can be solved. Technology has become the real religion for masses of people. The technological way of life means that people are free of taboos, immune to irrational authority, themselves reasonable and peaceable and tolerant of differences, progressive in outlook, and enjoying the abundance of the abundant life.[1]

The Corruption of a Dream

However, that euphoria has already peaked, sometime within our lifetime. Today that utopian vision begins to sound like black humor. Due to the eugenics experiments on living humans done by the Nazis and the Hiroshima burning of living humans done by Americans, due to the mad scramble for multiple warhead missiles, the worldwide nuclear radiation and the pollution of air and water, the prospect for a "drugged and brainwashed 1984," and the knowledge that the innovative cures for our troubles often produce unforeseen consequences worse than the original troubles—due to all this, the secular vision of utopia now seems phony, trivial, wasteful, and often, as Paul Goodman says, "a moral scandal."

[1] See Paul Goodman, "Can Technology Be Humane?" *New York Review*, November 20, 1969.

SOME PRO'S AND CON'S

Many sensitive writers are beginning to express grave doubts about the course of technology. The cornucopia begins to look like Pandora's box.[2] Even science fiction is pessimistic about the future.

What exactly are the charges that can be made against the technological spirit and its vast achievements?

1. *Emphasis upon Material Consumption.* Perhaps its major virtue is also its major fault: it makes material things, the production and consumption of goods, the be-all and end-all of living. Contrary to what Jesus said, technology prompts people to believe that life does consist in the abundance of one's possessions. It tends to define the good life as the life of many goods. The technological spirit seems motivated by the drive to produce and consume, and every new use of the machine increases the affluence which already dominates our common life. Obsession with prosperity may be America's form of mammon worship. Trusting in technology to deliver us from all our troubles, we are tempted to "worship and serve the creature rather than the Creator."

According to Archibald MacLeish,

[2] As an example of the serious criticisms being made, see Michael Harrington, *The Accidental Century* (New York: Macmillan, 1965), pp. 13-17.

PLENTY AND TROUBLE

We are ceasing to think of ourselves as men, as self-governing men, as proudly self-governing makers of a new nation, and are becoming instead a society of consumers: recipients—grateful recipients—of the blessings of technological society . . . It never crosses our minds apparently . . . that a population of consumers, though it may constitute an affluent society, can never compose a nation in the great, the human sense.[3]

The English were once said to be a nation of shopkeepers. We Americans, then, are a nation of consumers, and technology aggravates this human frailty.

Does this mean that Christian people must condemn technology and renounce all its benefits? It all depends on who benefits. If the continuous expansion of technology benefits primarily those who already have plenty and too much, then further expansion only increases human selfishness and adds to human sin. If it serves to make the rich richer and the poor poorer, then we must renounce it. But if the new technology can benefit the lowest 20% of our people to raise them out of poverty, if the machine can lift the developing nations, then it deserves Christian support.

2. *Tendency Toward Depersonalization.* A statement from the World Council of Churches

[3] "The Great American Frustration," *Saturday Review*, July 13, 1968, p. 15.

SOME PRO'S AND CON'S

puts forth another charge against the technological spirit.

> The chief problem of technological culture is that it contains something within itself which, in spite of its achievements and in part because of them, tends to establish the impersonal, the mechanical, the indiscriminate as that which is dominant. . . . The scientific method is irreversible and unpredictable in its long-range results. There is a dynamic in the march of scientific discovery and technological advance which defies the decisions of men.[4]

There seems to be something in the technological spirit that inclines us to understand all things, including other people, as machines. It seems to depersonalize the world. Everywhere it tends to erase the I-Thou experiences and replace them with I-It relations. The worker feels this when his job gets automated; the pupil feels it when he studies with a teaching machine; and the patient feels it when he walks into a lab for a physical check-up. The whole world feels stripped of people, and I, even I, begin to feel like a machine in a world of machines.

To some extent people have always been haunted by the fear that man-made devices might

[4] "Christians and the Prevention of War in an Atomic Age," World Council of Churches, Division of Studies. Published as a provisional study document, Geneva, 1961.

PLENTY AND TROUBLE

overwhelm and devour them. The stories about the sorcerer's apprentice who almost drowned his world, Frankenstein's monster who tortured his creator, and the science fiction androids that outperform human beings—all play upon the ancient fear that men's imaginative minds will conjure up demons too powerful to control. Among us the computer, the newest glamour machine, fires up our fears again. For some reason the new technology nurtures a suspicion and distrust of itself. Whether the facts justify the fear is not as important as the fact that such fears do get aroused. People *feel* dehumanized, and that is what counts. The sculptor Giacometti expresses it in his work, "The Captured Hand," a twisted hand grasping for freedom before the machine devours it.

3. *Technological Progress Is Always Ambiguous.* Prof. Mesthene, whose studies of technology are everywhere praised and widely quoted, points out that technological change has incalculable prospect for good and at the same time an inevitable by-product; it destroys as it creates. "New technology creates new opportunities for men and societies, and it also generates new problems.... It has both positive and negative effects, and it usually has the two *at the same time and in virtue of each other.*" [5] For example, modern plumbing

[5] Mesthene, *Technological Change*, p. 26.

SOME PRO'S AND CON'S

destroys the village pump, and with it the conversation, the friendships, the concern for neighbor which flourished around the public well in the town square. Technology puts a car in every driveway, but many repair shops have a sophisticated new testing device (a kind of stethoscope, electrocardiograph and electroencephalograph combination for ailing cars!) which deprives the mechanic of his skill and the pride that went into his earlier work.

Many human functions waste away under the impact of technology. Most laborers do not "labor" these days. They operate machines by pushing levers and buttons, so they organize bowling clubs for exercise! Alice Mary Hilton asks a valid question.

When man no longer needs to pull the plow and fell the trees and forge the iron, he must find other tasks to satisfy his restless nature. How will he tire his muscles to earn his rest? How will he use his mind to find his peace? How will he stand upon the earth he has not tilled in the sweat of his brow, and feel that he is its master? What will man do with his life when he no longer has to labor to earn his right to live? [6]

The skills and crafts that once gave meaningful work to creative men now are confined to hobbies.

---

[6] "Cyberculture in the Transition from a War to a Peacetime Economy," *Fellowship*, May, 1964.

PLENTY AND TROUBLE

When a person forfeits his pride in work, and loses his sense of competence, and feels inferior to the machine he operates because it can work faster, longer, and more accurately than he can, then he loses the image of himself as a creator. Mesthene is right. Every new use of the machine destroys something of value.

Jacques Ellul has written an extended note on the theme, "technical progress is always ambiguous," in which he argues that contradictory elements of good and bad "are always indissolubly connected."[7] Ellul makes four points:

1) All technical progress exacts a price. For instance, due to new techniques in medicine, people live longer, but at the same time "life has become very much more precarious; our general state of health has become very much more fragile.... Thus, though we live longer, we live a reduced life with nothing resembling the vital energy of our ancestors."

2) Technique raises more problems than it solves. For instance, new techniques of public health have resulted in a population explosion which has produced large scale hunger and the prospects of massive famine.

3) Evil effects of technique are inseparable from the good. All mechanical progress, for instance,

---

[7] "The Technological Order," quoted in Stover (ed.), *The Technological Order*, pp. 28-37.

SOME PRO'S AND CON'S

necessarily makes for unemployment, at least *interim* unemployment for millions of workers who thereby pay the price for social progress. Also, the new techniques have "greatly reduced the trouble, the difficulties, and the anguish implied in killing people. A bombardier has no feeling of killing anyone. . . . In such ways, then, positive elements of [technology] result essentially [by very complex expedients] in favoring war and even in provoking it, even if no one has the *intention* of using technique 'badly.'"

4) All technical progress contains unforeseeable effects. In this case, experiments with drugs for treatment of human disease and the use of DDT for control of insects are familiar illustrations.

Thus, Ellul and Mesthene appear to agree that technology carries with it, *by its very nature,* danger, increased risk, and some outright damaging consequences.

4. The Strain of Living. If technology destroys something of value, it often also adds something of trouble. It makes living considerably more complicated. Let me add up several kinds of trouble, including ulcers, heart attacks, "nerves," psychic disorders, suicide and violence, and give them the gentle name of "strain," and then say that technology adds to the strain of living. TV at dinner time brings news of wars and disasters that tug on the

PLENTY AND TROUBLE

conscience and make dinner harder to digest. Driving a car at sixty-five miles per hour is harder on the nerves and considerably more dangerous to the body than driving a one horse shay. Old Dobbin at least got you home safely from the tavern, however drunk you were.

The new medical technology will confront people with troubling questions. They will have to choose the sex of their unborn children, and when their aged parents in terminal illness depend on further medication and machine sustenance, sons and daughters will have to decide when to let them die. Prof. Norbert Wiener, known as the father of cybernetics, asks a sobering question about the patient who knows that his doctor has to make decisions about such use of technology. "What if every patient comes to regard every doctor not only as his savior but as his ultimate executioner? Can the doctor survive this power of good and evil that will be thrust upon him? Can mankind itself survive this new order of things?" [8]

Technology offers more choices, and makes choosing more difficult. In moral questions, for instance. Kenneth Boulding points out that the seven deadly sins may not be deadly any more. There will be no occasion for envy when everyone can have all he wants. He can indulge in

[8] *God and Golem, Inc.*, p. 68.

SOME PRO'S AND CON'S

gluttony without getting fat, in sloth without getting poor, in anger without getting into trouble, and lust without fear of pregnancy. The fear of earthly consequences has disappeared, and "in the developed society it is the pure heart or nothing." [9] But can we be virtuous without the prodding of punishment for our sins?

Such are the new strains we confront because of technology. It upsets the routines, contradicts old habits, requires constant reappraisal of morals, requires that we abandon many familiar ways of behavior, and insists that we experiment with change. It makes life a turmoil, and fascination with the new often gets outweighed by threat in the new. How much novelty can a person manage? Even the body cannot "take" everything. When a person travels by jet plane across several time zones, it requires some days before his body can adjust to the changed rhythm. How much more true it is of his mind, his temperament, his political structures, his whole tempo of life! How much new experience can one man or one society digest at one time? Perhaps the hidden and unrecognized consequences of technological change account for more of our present troubles than we have yet assumed.

[9] "The Wisdom of Man and the Wisdom of God," *Human Values on the Spaceship Earth* (National Council of Churches, 1966), p. 13.

PLENTY AND TROUBLE

5. *Encouragement to Pride.* Prof. Mesthene says that our technical prowess literally bursts with the promise of new freedom, enhanced human dignity, and unfettered aspiration. He believes we have now, or know how to acquire, the technical capacity to do very nearly everything we want. Can we transplant human hearts, control personality, order the weather that suits us, travel to Mars or Venus? Of course we can, if not now or in 5 or 10 years, then certainly in 25 or 50 or 100. And a medical worker said it sharply, "We cannot duplicate God's work, but we can come very close."

All this constitutes a temptation to human pride. The advancing technology seems to abolish all mystery and "to make the blasphemous promise of solving *all* human problems." (Deats) For instance, a doctor said that once we solve the problems of cancer and arteriosclerosis, then except for murders and accidents, men will stop dying! Serious scientists are serious about this. They experiment with cryonics, the deep freezing of living tissue in the hope of thawing it back to warm life again. Such experiments justify the theologian in saying,

What blurs the presence of God, first in the human imagination and later in reflective thought, is the unveiling of these secrets and the controlling of these

SOME PRO'S AND CON'S

forces.... Technology involves man frankly in the making of the universe. Man assumes this power of world-making; he gives the universe its meaning, and in it he experiences and affirms his own autonomy. ... The result of technology, then, is a true revolution in the imaginative and mental habits of people.... The very nature of technology ... tends to banish all sense of mystery.[10]

And perhaps all sense of humility.[11] Technologists tend to feel that everything conceived in the name of technology must be good, but the World Council of Churches warns against any such faith "in the capacity of education or technology to solve all human problems ... and the failure to recognize the imperfect, precarious, and transient character of all human achievements." At the New Delhi Assembly the Council declared, "It is not good that man should be subdued by nature or enslaved by technology. Nor is it good that nature should be mastered by man, if the mastery merely feeds his rebellious pride. But it *is* good that man should increase his knowledge

---

[10] Marie Dominique Chenu, *Faith and Theology* (Dublin and Sydney: Gill and Son, 1968), trans. Denis Hickey, pp. 220-21.

[11] It is only science fiction, but Arthur C. Clarke puts it vividly in *2001: A Space Odyssey*. "Man's control over the machines is absolute. He has manipulated his natural environment, conquered the problems of interplanetary travel and is ready for what comes next, in the first year of the twenty-first century, 2001."

PLENTY AND TROUBLE

and should use his growing mastery of nature for the benefit of mankind and the glory of God."

St. Paul warned the Romans they were misusing their knowledge of the natural world. They should see the Creator in all that was visible; instead they were worshiping the work of their own hands. (Romans 1:18-25) The machine, indeed, tempts men to such mistakes; they incline to build a tower of Babel, to usurp the powers that belong to God. Technology is most fascinating, therefore most dangerous.[12]

The Quality of Life

In order to take the measure of technological change, we must notice also that many consequences are unforeseen. A partial list of the by-products spun off from technological change will suggest what I mean: the population explosion itself, which has brought crowding, waiting in lines, and closed doors to young people applying to college; traffic jams, with wasted time and frayed nerves; pollution of air and water; atrophy of physical health due to riding in cars, breathing the air and eating chemically treated foods; atrophy of the power to entertain ourselves due

[12] Among the theologians most critical of technology is Emil Brunner. See for example his *Christianity and Civilization* (New York: Scribners, 1949) II: 4-5, 13.

SOME PRO'S AND CON'S

to dependence on TV and the movies; radiation damage from bomb testing; tedious work on machines; the aggravation of social injustice when the poor get poorer; the concentration of power in the hands of mass media; the data bank run by government; junk mail and trash magazines and useless advertisements mailed to "the occupant," and the 500-page Sunday newspaper that takes until Thursday to read. All such consequences of technology are unintended, unforeseen, and, what is most serious, unforeseeable! Human beings assume unpredictable risks when they launch out on the technological venture. Who can imagine the consequences of controlling the weather, for example? Human nature is flexible indeed, but fragile. Sensitive but delicate. It remains to be seen how human nature can respond to unforeseeable consequences.

The changes that occur in our common lives under the steady influence of technology make a subtle transformation in the quality of our living, and that quality can degenerate so unnoticeably that, before we know it, our familiar way of life is no longer recognizable. As we consume more things we become dependent on them, alas to our to our embarrassment when suddenly those things are threatened. As our social world expands, the number of our acquaintances increases but each relationship becomes more shallow. As we increase

PLENTY AND TROUBLE

our communal power to manage life, our individual power over our destiny gets whittled away until we feel powerless to shape our own careers. As we become conscious of more people and more events, we lose our privacy, our breathing room and living space. Something about the quality of life gets threatened. Socrates asked about the value of shipbuilding if men don't know how to navigate. I wonder about the value of technology if the purpose of living is not clear.

All these charges make up a serious case against the utopian dream of a technological world. They need to be scrutinized carefully as we come finally now to attempt a forthright Christian statement of principles.

7

A Christian Critique

The hurried inspection in the previous chapter of the pro's and con's of technological change gives a divided report. As with double vision, we cannot bring it into focus. Technology brings enormous benefits through the mastery of nature, and at the same time it destroys nature. It relieves human toil, but destroys human skills. It cures disease and cuts the death rate, thereby making a population growth that brings crowding, famine, and starvation. As communication, it draws people together in peace, as weapons it threatens to abolish mankind. It makes people comfortable, and avaricious. On and on goes this list, describing technology as apples and snakes, both! Men wonder whether the eschaton is to be a "far-off divine event toward which the whole creation moves," or the apocalypse. Will 1984 see John's

PLENTY AND TROUBLE

"new heaven and a new earth" or Huxley's "brave new world"?

When you set aside the "hardware" and reflect about the mood spirit of technology, again you reach an ambivalent judgment. On one side you sense the dynamic, buoyant, achieving spirit of technology. Whatever good thing needs to be done can be done! No longer are men confined to the accidents of nature and the fate of their own limited powers; now they make their own destinies and shape their own futures. When the astronauts first orbited the moon, Archibald MacLeish reflected,

The medieval notion of the earth put man at the center of everything. The nuclear notion of the earth put him nowhere, lost in absurdity and war. The latest notion, formed in the minds of heroic voyagers, may remake our image of mankind. No longer that preposterous figure at the center, no longer that degraded and degrading victim off at the margins of reality and blind with blood, man may at last become himself. To see the earth as it truly is, small and blue and beautiful in that eternal silence where it floats, is to see ourselves as riders on the earth together, brothers on that bright loveliness in the eternal cold— brothers who know they are truly brothers.[1]

[1] Quoted in an editorial in the *Boston Globe*, December 28, 1968.

A CHRISTIAN CRITIQUE

Against that rhapsody there sounds the thunder that makes us doubt the dawn. Everything overhangs with ominous threat, a strange disquiet. The ground underfoot feels like an earthquake in process; *terra non firma.* Huge movements of population, the natural damage, the medical tampering, the police controls begin to feel like an irresistible shaking of the foundations, too threatening to be controlled, too complex to be understood. Again, the technological experience feels contradictory.

But to move beyond this stalemate in our understanding, to say something more than Yes *and* No, let us try finally to enumerate a few Christian principles. The five I use are not Christo-centric, and they do not emerge out of any elaborate ethical system, yet they seem so elemental in the Christian understanding of things and so relevant to the technological problem that they serve to outline our Christian critique of this matter.

1. God Meets Us in Our Strength

If we are honest we must confess that technology has eroded the religious spirit. For two hundred years Christianity has been bombarded by the technological spirit. During these years men have really come to know that they are competent to handle their own affairs. Once they prayed for

PLENTY AND TROUBLE

rain, but now they entice rain from the skies and desalinize the ocean water. Once they prayed for recovery from disease, now they hurry to the hospital for computerized exams and microscopic tests. Men and women manage their own affairs, fortified by the conviction that they are but "little less than God" and given "dominion over the works of thy hands, for thou hast put all things under (our) feet." They feel fortified by Bonhoeffer, who made bold to say "God is teaching us we can get along very well without him," and by Bishop Wickham, who believes God is driving us "to attain a maturer religion that relates God to the things where man is strong, rather than restricting Him to the shrinking areas in which man is weak." This is heady stuff, indeed, but it takes seriously the understanding that God compels men to fashion their own destinies. "Faith moves mountains," said John Calvin, when "armed with axe and spade." Faith moves plagues when armed with DDT and hidden sewers. Faith removes drudgery, armed with computer and the hydraulic press.

Lord, shall we not bring these gifts to Your service,
Shall we not bring to Your service all our powers
For life, for dignity, grace and order,
and intellectual pleasure of the senses?

A CHRISTIAN CRITIQUE

> The Lord who created must wish us to create
> And employ our creation again in His service.[2]

The Christian person therefore says Yes to technology because, through the technological process, God enables him to meet his neighbor's need effectively. The Christian cannot join the Luddites. Luddites roamed the early nineteenth century English towns smashing the new textile machines, and quill penmen rioted when the printing press came to Paris. In our day the firemen who ride the diesel engines and the Linotype operators who insist on handsetting are the symbolic Luddites, trying to hold off the new technology to protect their ancient jobs. Featherbedding is a gentle form of rioting. Stop automation, stop change! they cry, but stoppers always lose in the end.

Sensible men are now aware that technology cannot be erased. We are not *becoming* technological, we already *are* technological people. God meets us in this new situation, as in times past he spoke from the burning bush and the Sinai thunder. The older conceptions of destiny, fate, human limitations, surrendering to the unfathomable will of the Almighty—all such submission is now replaced by the experience of hope and ac-

---

[2] T. S. Eliot, Choruses from "The Rock," *The Complete Poems and Plays of T. S. Eliot* (New York: Harcourt Brace Jovanovich, Inc., 1962). Quoted by permission of Harcourt Brace Jovanovich, Inc., and by Faber and Faber Ltd., London.

PLENTY AND TROUBLE

tion and confidence which define the self-image of technological people. We are no longer slaves in Egypt. We have moved out toward the New Land, and there God meets us in our inventiveness and in our refusal to be satisfied with old remedies. We do not honor God in our weakness any more than we honor him in our sin. We glorify God with our virtue, virtue of mind and muscle. We find God where we solve problems more than where we are baffled. We serve God where we achieve, more than where we are defeated.

2. We Are Chosen for Responsibility

Heady stuff, I repeat. People know God has given them dominion over all creation, and that dominion, expressed in the spectacular new technology, entices some people to boast, We are equal to God! and still others to claim, We have replaced God! However, there are two ways to read that verse in Psalm 8. One emphasizes,

Thou hast made him but a *little less than God*
And given him *dominion over the works* of thy hands.

Or it can be read,

Thou hast made him but a little less than God
Thou hast given him dominion. . . .

By this reading the human person who holds these

A CHRISTIAN CRITIQUE

vast powers received them from the Creator who created. The person is created by the One who put creation under his feet. That makes the man accountable. He holds powers in custody.

God no longer allows us to be adolescent. He has sharpened up for us the ancient choice of life or death, maturity or disaster, responsibility or ruin. The danger appears apocalyptic. Technological powers make possible thoughtless destruction. In late summer, 1970, the army had to go through a legal dispute over the burying of nerve gas rockets in a watery grave off the Florida coast. Men had created something they were unable to live with and apparently unable to destroy in any safe way! Technology makes also for incredible self-deceit; after all, the army meant the nerve gas to threaten other people, not Americans! Such childish behavior is no longer permitted us. We cannot play with creation; its powers are too dangerous. We are responsible for nature. The second great commandment must now be modified, "You shall love your neighbor as yourself and care for creation as your home." Once we were instructed to love other people because God is incarnate in those people. Now we are instructed to care for nature because God gives the good earth as a home for his people. As God's people we are responsible for God's good earth.

This requires that we take charge of things with

PLENTY AND TROUBLE

energy and rational oversight. That includes stringent political controls over the urban sprawl, poisoned waters, crowded ghettos, mechanized medicine, computerized business. There is a great deal of profit motive behind the pollution of nature, and slothful taking-the-easy-way-out. Businesses replace men with computers and hospitals replace nurses with machines because it is the efficient thing to do, and who will argue with efficiency and prosperity? Both citizens and politicians too often play Brer Rabbit; they "lay low and say nothin'."

Responsible social controls will aim first at what we can call the technological syllogism. The general notion prevails that whatever technology can do must be done. If we can get to Mars, we must do it. If we can build the SST, we must build it. That way, capacity determines program, technology becomes teleology. Or, in McLuhan's fashionable slogan, the medium is the message. Call it the technological syllogism. Major premise: Whatever we want to do we can do. Minor premise: Whatever we can do we ought to do. Conclusion: Whatever we want we ought to have. That logic equals moral and intellectual abdication to the technological spirit. "Whatever is possible is necessary" means we capitulate to the clever technicians who conjure up all sorts of devices, demonic or wasteful or trivial.

A CHRISTIAN CRITIQUE

To resist that syllogism, some hippies take to the rural commune, other people take to drugs, still others engage in strange spiritual quests, but responsible people will take up their political tasks and express their concern through vigorous public action. Technological decisions are too important to leave to technologists and managers. Already the technical people dominate public decisions because they can promise exact and immediate results, and deliver! We have been to the moon, they argue, therefore we can get to Mars. We have carried a hundred passengers in jet planes, therefore we can carry five hundred in supersonic planes. We have built nuclear subs, therefore we can build a multi-headed atomic missile. But whether it is worth going to Mars, worth packing half a thousand people into one enormous plane, or worth blasting ten cities with one shot—those questions belong to the people. Technological achievements don't have to be done simply because they can be done, but they will be done unless responsible people begin to say No. Controls over technology must arise from the people, else the garden will have more snakes than apples.

3. The Supreme Human Trait Is Valuing

The distinctive function of the human person is not to make tools (*homo faber*) nor to think

PLENTY AND TROUBLE

(*homo sapiens*) but to make judgments, to evaluate. The word "value" sounds abstract as though it belongs to the philosopher's vocabulary, but to value means to decide. Solomon prayed for wisdom "to discern between good and evil." To say Yes or No, to say I will or I won't, to like and to dislike—that is valuing. Whenever a person chooses sides on some substantial question, that makes him a man.

Some values clearly are more Christian than others. To choose the universal rather than the provincial, for example. Provincial values center around wealth, race, class, nation (read affluent, white, middle, American) and these obviously do not rank with the Christian understanding that all men equally belong to God. Again, according to the prophets and the Magnificat and the Last Judgment parable, it is a Christian value to protect the dispossessed and the underprivileged. Such values are conservative; they endure for a long time and are slow to change. By their inherent nature as values, they are stable, to be vindicated in the long run.

Yet values do change, and in a society of rapid change values change rapidly. Technology has upset, for example, the traditional values associated with rural life, handcrafts, individual enterprise, privacy, and the like. Beyond these specific changes there are two fundamental questions that

A CHRISTIAN CRITIQUE

Christians need to ask about the values of technology.

First, we must question the technological assumption that man's main problems are technical problems, which have only technical solutions. That is a value judgment, and it deserves severe questioning. The technological society assumes that technology can resolve most human problems. A panel member at the AAAS meeting, for instance, said that as soon as the biomedical problems get resolved, the theological problems will dissolve away. True, men confront severe technical problems, such as overcrowding, disease, drudgery, pollution, and we must rely on the evolving capacity of technology to help find "solutions" to the problems technology itself has created.

But notice. Every such "solution" reinforces the impact of technology in its totality, which includes more and more unforeseen problems arising from the new solutions. Human survival does require more technology, but every struggling step out of the engulfing spirit of technology only bogs us down further into the quagmire. "The further we advance into the technological society, the more convinced we become that, in any sphere whatever, there are nothing but technical problems" (Ellul) and technical solutions. A good friend of mine, a thoughtful person whom I great-

PLENTY AND TROUBLE

ly respect, recently fell into this trap by writing, "The dimensions of the danger are so great that only a program equally vast in scope can hope to deal with it. The same technology that has brought us to our present plight can, if properly used, save us from it." Then he pointed to the spectacular things technology can do, and concluded, "Used according to the needs of various cultures, technology can help accomplish all these things without the depersonalizing effects so noticeable now." In reply to that I see this Christian question: Can we rely on technology to solve the problems it has created when one of its basic assumptions is that it deals with the real problems and it alone has real solutions? Theodore Roszak may be closer to the truth, that "the great secret of the technocracy lies in its capacity to convince us (contrary to everything the great souls in history have told us) that the vital needs of men are purely technical in character." *That* value, deeply embedded in technology itself, constitutes one overwhelming value which Christian people must resist. They know there are other problems more critical and other solutions more humane.

The second question Christian people put to the technological spirit arises out of the question whether technology is fundamentally an instrumental tool or a determining process. A good deal can be said to justify the belief that it is only a

A CHRISTIAN CRITIQUE

tool, to be used for whatever purposes its managers decide. After all, the computer eases human drudgery and hi-fi equipment brings Heifitz to the living room. Technology serves human values! However, there is the danger that technology becomes so efficient, so attractive, so outpouring of its benefits that only the efficient and attractive and abundant things appeal to people. People begin to want what they can have easily, and that makes technology a frightening thing to behold. I begin to share MacLeish's numb, unformed feeling that we human beings "have somehow lost control over the management of our human affairs . . . the process itself has somehow taken over, leaving the purpose to shift for itself so that we, the ostensible managers of the process, are merely its beneficiaries." Or its victims. With Hiroshima it became clear for the blindest man to see that technology is loyal not to humanity but to truth, its own truth, and its truth is not the law of the good but the law of the possible. Whatever is possible becomes necessary, and *that value* carries the essential scourge of the technological spirit.

At two fundamental points, then, the Christian has to question the values of technology: Does technology emphasize the technical dimensions of life's problems so drastically that it neglects or denies the nontechnical nature of human nature?

PLENTY AND TROUBLE

Secondly, does technology lose its instrumental position and become the determiner of events according to its own logic?

Having raised those questions, and never forgetting them, we need now to acknowledge the contributions which technology makes to the valuing process.

"Technology has come of age," says Prof. Mesthene.

We have the power to create new possibilities, and the will to do so. By creating new possibilities, we give ourselves more choices. With more choices we have more opportunities. With more opportunities we can have more freedom, and with more freedom we can be more human. That, I think, is what is new about our age. . . . Our technical prowess literally bursts with the promise of new freedom, enhanced human dignity, and unfettered aspiration.[3]

In a profound sense, technology does heighten the valuing process. It exposes the ethical issues with new clarity and urgency. The Minister of Technology in Great Britain believes that when the computer can calculate the results of any given policy and foresee its consequences accurately, then the debate need not concern the pru-

[3] Emmanual G. Mesthene, "What Modern Science Offers the Church," *Saturday Review*, Nov. 19, 1966, p. 30; based on his address to the Conference on Church and Society, Geneva, 1966.

A CHRISTIAN CRITIQUE

dential questions but can concentrate on the value issues. Men will not argue about the possible consequences of a given decision, but will decide the question on its merits and ask only one question: Is this policy good for people?[4] In the previous chapter we noted how Kenneth Boulding observed that technology requires a more mature morality. If a person can indulge in gluttony without getting fat, in sloth without getting poor, in lust without fear of pregnancy, then indeed "in the developed society it is pure heart or nothing." Clearly, the advancing technology heightens the valuing process.

Therefore, we have to be cautious about condemning technology as a destroyer of values. Of values, yes; of valuing, no. It is easy to confuse values with stability. Values change, but "the human activity of valuing, and the social functions of values, do not change. That is the source of the stability necessary to moral experience." In the days ahead, Mesthene continues,

the emphasis will have to shift from stable, known, and familiar values to the conditions for and mechanisms of valuing. For it is not existing values as such that are valuable, but the continuing human ability to extract

[4] See A. W. Benn, "Living with Technological Change," *New Statesman*, Dec. 13, 1968.

PLENTY AND TROUBLE

values from a changing experience and to use and cherish them.

That value, I would argue, is in no way threatened by technology.[5]

People do the valuing, and they must hold the valuing function in their own hands, including their judgments about whether technology enhances the valuing process or destroys it by its own dominating and consuming nature.

4. Caring About People Counts the Most

Hiroshima made it clear (so do SST, MIRV, *in vitro*, the Army's nerve gas, and the hospital's computerized exam) that technology is committed not to people and to human good, but to truth—its truth that whatever is possible should be done. *That value* constitutes a demonic element in technology.

Unhappily, we human people sell our souls for the juicy stew that technology serves up. It looks like a good bargain, and our appetite for the benefits deadens us to the compromising cost. Almost everything we eat, wear, or use comes through the technological process, treated chemically, controlled electronically, handled, counted,

[5] "Technology and Human Values," *Science Journal*, Vol. 5A, October 1969, pp. 46-47.

A CHRISTIAN CRITIQUE

packaged, delivered to us by the anonymous mechanical devices. Our life depends on a technology we do not comprehend and cannot control. We accept the dangers as a matter of expedience.

I accept the noise, dirt, ugliness, the crowds, the dehumanization and all this absurdity in order to have my comforts, my medicine, my automobile, library, and orange juice. I doubt that mankind can tolerate our absurd way of life much longer without losing what is best in humanness. Western man will either choose a new society or a new society will abolish him.[6]

Our time is rightly called the technological society, not because of the huge abundance of computers, rockets, electric-eye doors and genetic experiments, but rather because we live our lives as recipients gladly accepting its benefits rather than as controllers soberly expressing some grander purpose. We are not in control. The process controls us. The technological spirit saturates the human climate and rules our life.

The computer which handles only quantifiable data gives aid and comfort to those who think all the important things in life can be learned by looking at quantifiable data. "The real danger is not that machines will begin to think like men,

[6] Dubos, *So Human an Animal*, p. 195.

but that men will begin to think like machines." (Sydney J. Harris) We feel manipulated; indeed we are manipulated, by a people-manipulating technology which began with the megamachine of the Pharaoh's slave system in Egypt, ripened in the Benedictine monasteries of Europe,[7] and climaxed in the machinery of contemporary life: the political machine, the military machine, the machinery of government, the corporation machine. We are manipulated and we get accustomed to it because the technological spirit renders itself invisible. If the Devil's best trick is to persuade us that he does not exist, so technology's chief success consists in its capacity to make its assumptions as unobtrusive and as pervasive as the air we breathe. Technology assumes a position somewhat like that of the baseball umpire. Normally he is the least conspicuous man on the field, yet he may be the most influential, because he makes the rules, he keeps the score, he judges the players. So does technology, too often.

Generally speaking, we do not consider technology as an optional thing. Rather, we accept it as inevitable. It assumes the place of a grand cultural imperative beyond debate. Take a common example, the automatic answering service.

[7] See Mumford, *The Myth of the Machine,* pp. 194, 264.

A CHRISTIAN CRITIQUE

You telephone a doctor or museum or business office, and the automated tape answers you with an oh-so-sweet voice. "So-and-so is busy at the moment. If you want something urgent, speak a message here and he will call you back. Meanwhile, here is some other information you didn't ask for. Good-bye now." We put up with this indignity because it is efficient. It releases some telephone hostess to do more productive work elsewhere, and anyhow the machine is cheaper than the girl. There you have it! Efficiency, cheapness, productivity, those are the high gods we worship, and in their honor we endure all sorts of insults to our dignity. By accepting them we become technological people. We sell out to the demonic spirit of our time.

Basically this spirit is inhumane, and that constitutes its major defiance of the Christian ethic. Concern for people is the elemental Christian virtue. Caring, compassion, concern, love—whatever you name it—is the basic moral command. Public protests, student revolts, anti-war rallies, welfare mothers' marches, demonstrations against corporations and universities—shortsighted, mistaken, violent, mixed in motive as they may be—boil down to a struggle to be humane. In a world that feels hard and calloused, a world that begins with commercials and ends in Vietnam and the ghetto, young people (young in heart, of all ages!)

PLENTY AND TROUBLE

are determined to reassert the (unrecognized) Christian principle that people count more than machines and efficiency. Caring about people is the only thing that makes a man a man. Technology must knuckle down under that principle, and if, because of its very nature, it cannot accept that principle, then technology is demonic by definition.

I am saying something about people, basically. I do not accept Descartes' definition, *Cogito, ergo sum* (I think, therefore I am.). I believe rather, *Sentio, ergo sum.* I feel, therefore I am. That simple act of feeling and caring about other people is so elemental we are ashamed of it. It embarrasses us technological people.

5. Life Consists of More Than Abundance

It is easy to assume, as the free enterprise economy has indoctrinated us all, that the bigger the better. If the GNP (gross national product of goods and services) only expands faster than the population rate, that makes prosperity and welfare for all. That doctrine simplifies the problem of living into the aphorism, The Good consists of many goods. In clear defiance of the gospel it says that life does consist in the abundance of one's possessions, and the technological dynamo

A CHRISTIAN CRITIQUE

continues to expand our producing and consuming.

Yet those who reflect soberly about the course of human events are beginning to question this excessive concern for growth of the GNP. Udall calls it a "destructive gospel." The Chamber of Commerce syndrome that growth is always a good thing deserves a serious critique.

> Taught to believe that what is big is blest,
> We build and fell and drill and strip and mine.
> We gorge and glut the land from east to west,
> Not sensing that the cancer is malign.[8]

This growthmania endangers human life, depletes the natural resources, pollutes the earth, piles up junk, threatens human health, and makes the environment ugly and frustrating. A new skyscraper or highway often causes more pollution, congestion, and crime than it can outweigh in new jobs and taxes. By producing too much, we destroy what we have!

After the fundamental needs of people have been met, further growth in GNP majors in trivia. For poor nations and the poor sections of wealthy nations, a rising GNP means food, better homes, better education and basic health, but for rich

---

[8] Edith Lovejoy Pierce, *The Christian Century* (April 15, 1970), p. 438. Used by permission of the author.

PLENTY AND TROUBLE

nations and the affluent majority it means trivia: electric toothbrushes, power windows in the car, and ten brands of deodorant. Does further growth of the GNP in America really make people happier, or does it make them more miserable, with its insulting avalanche of trivia we do not need and never thought of wanting until someone marketed it because it was possible to do so?

If we are to achieve what Galbraith calls the "economy of beauty" we will need to revise our definition of the GNP. It is not good enough to measure prosperity by the cash value of the goods we produce and the services we perform. Goods do not add up to The Good. There must be beauty, too, and space to move in, and quiet to enjoy, and cleanness in the parks and sanitation in the rivers and purity in the air, and health for the birds of the air and the fish of the sea. Such intangibles and invisibles belong to our human lives, and we are not living abundantly unless we have these things along with our bread and butter. T. S. Eliot asked, "Where is the life we have lost in the living?" Denis Gabor suggests a reply.

The epoch of dynamic man, after two successful centuries, must somehow come to an end, and we must approach the state of maturity. . . . Man reaches maturity when he stops growing in height and weight, but he need not stagnate; he can start developing new

A CHRISTIAN CRITIQUE

faculties. This is what humanity must do too; stop quantitative growth and start growing in quality.[9]

Consequently, the economists talk seriously about developing a steady state economy. That attempt raises serious economic problems in a free economy, but many people are impressed by the fear that unceasing growth endangers quality of life. The Greeks felt size incompatible with excellence, and Americans are beginning to wonder.

A steady state economy would require that people live more simply. Scott Paradise suggests that the automobile itself may be a pollutant. Not the exhaust—that can be remedied by control devices, or by using a gas turbine or battery-powered engine. Not the exhaust, but the car itself may be the pollutant. The car makes noise, consumes space, and turns the city into a madhouse. It kills fifty-five thousand people a year, and wastes enormous amounts of time in traffic jams. It piles up in ugly junkyards, and worst of all, it creates an economy dependent upon it, and that constitutes tyranny. The private automobile pollutes the quality of life. We may have to live without our private cars, and without many other things that make for waste, such as detergents, and food disposals, and seven-gallons-of-water-

[9] "The Future of Industrial Civilization," *Student World*, no. 2, 1968, p. 159.

PLENTY AND TROUBLE

per-flush toilets, and countless electrical gadgets, and the packaging that consumes enormous amounts of paper, hence wood, hence forests. The new style of living would be frugal, sparing, cautious, niggardly, conservative—all those unpatriotic and uncapitalistic qualities which may nevertheless just happen to be humane and Christian. We must find how to contradict Jules Feiffer's jibe that pollution is as American as apple pie. Pollution is someone's profit and everyone's cost, and it destroys what God has created.

The new asceticism, what Boulding calls parsimony, will include careful use of the natural world and a nonavaricious attitude toward possessions. It will emphasize quality of life. People will learn to enjoy rather than to possess. Beauty and solitude and silence will be cherished. Wildlife and wild scenery, with the smells and sights and sounds of the good earth will become joys again. People will not confine their delights to measuring and counting and analyzing, but will turn to wonder and amazement, to fantasy and festivity to round out their lives. They will try to be whole people again.

As a parable of our time, consider the moon probe. The Apollo spacecraft illustrates in microcosm the experience of living in a technological society. The Apollo is probably the most complex and sophisticated machine ever built. It generates

A CHRISTIAN CRITIQUE

its own electricity, it transmits television, its computer sends eighty billion bits of information back to earth daily, it navigates with great precision, and, despite its tremendous velocity, its speed and location can be detected with incredible accuracy at the distance of the moon. Yet as a living place it is archaic. It contains no civilized toilet; chlorination gives the drinking water a bad taste; freeze-dried food must be injected with water, kneaded in a plastic bag, then squeezed into the mouth; the dust won't settle; and the hammocks won't sag enough to be comfortable; there is no space to exercise, let alone move around in. The Apollo is a technical miracle and a living nightmare, thus a parable of our time. Our whole style of life seems miserable when contrasted with the high perfection of our technology.

Quality living means restraint with a purpose. The new asceticism does not aim to torment the body for the sake of the soul, but aims to restrain the appetites in order to share with the needy, to reject the trivial and to concentrate on the worthwhile. In personal life it expects the Christian man to possess as though he did not need to possess, and to be deprived as though he were not hurt by any deprivation. In his public life the Christian man will never imply that the poor, to be free of materialism, must be content with a low level of consumption. If he favors a rising

PLENTY AND TROUBLE

GNP, he must be sure he is concerned for the just distribution of goods that will raise the lowest people out of poverty. If he favors a steady state economy, he must be sure it does not perpetuate the gap between rich and poor.

Let the last word on this matter come from a secular historian, Lynn White, Jr. "Since the roots of our trouble are so largely religious, the remedy must also be essentially religious, whether we call it that or not. We must rethink and refeel our nature and destiny."

Utopia or Disaster?

At the beginning of each Christian year we are reminded that the whole Christian view looks forward to the future with hope. That theme has become the faddish style in the newest theology of Moltmann, Braaten, and others. We live essentially in the season of Advent. We look ahead to the coming of the Messiah. The prophets tell us to prepare, to watch, to wait, and to hope for the great event yet to come. The ancient dream for justice and peace will be fulfilled. Men will build houses and inhabit them, plant vineyards and eat the fruit thereof, and none will make them afraid. The Blessed One will come to usher in a New Time. He will cast down the mighty from their seats and exalt those of low degree; he will fill the hungry with good things and send the rich away empty-handed. That Christian view of the future, full of hope and huge promise—how different is it from the future promised by technology?

There is a technological view that sees only apples in the garden. Over a century ago the editor of the *United States Review* foresaw the time when "machinery will perform all work—automata will direct them. The only task of the human race will be to make love, study, and be happy." In our own time Dr. Simon Ramo believes that

production, distribution, transportation, communication, banking, engineering, law, medicine, education, government—all will, in the year 2000, be operated by new man-machine networks with all the men and information and material and equipment at the right place at the right time and told what to do and where to be, so the whole system will mesh and click according to plan.[10]

Another technologist, the president of a major oil company, declares, "If we can somehow maintain the proper path, the age of automation can become the Second Age of Pericles—an outburst of creative individuality in which a society based

[10] *Dateline*, October, 1967.

PLENTY AND TROUBLE

on the labor of the machine reaches out as far as the human spirit is capable." [11]

If all this comes to pass, it will be, according to Charles Malik, "a dreary and boring world where there is nothing beyond man and his mastery over nature, including his mastery over other technicians through his scientific management of them. Perfect hierarchy, perfect organization, total efficiency; but no spirit . . . no freedom, no joy, no humor, and therefore, no man." But it may not come to pass. Rexford G. Tugwell sees the year 2000 as the time when compulsory schooling will be extended to age 23 and retirement set at 50. There will be huge army and police establishments and vast investments in leisure and welfare programs to equalize things—and all this will have "finally exhausted the gains from technological advance." Did not the World Council of Churches remind us of "the imperfect, precarious, and transient character of all human achievements"?

What will the future be, utopia or disaster? Both, no doubt.

We live in a new time. Every generation feels that way about itself, but this time it's true! Only superlatives can get at the facts, only exaggerations are true. Never before in the long history

[11] Charles F. Jones, *Dateline,* April, 1970.

A CHRISTIAN CRITIQUE

of human affairs has mankind had to make so many profound decisions in so short a time. That is the overwhelming fact about our technological time. Never before has a great society been within grasp. Never has disaster been so threatening. Never have problems been as complex and solutions as baffling. Never has a new society been so desperately needed. Never have the alternatives been as striking.

Heraclitus had a *theory* of change, we have the *experience*.[12] He speculated that change is the only reality, that everything changes except change. But *we know* that no one will ever again live in the world he was born into, nor die in the world he lived in. Things are different, drastically different, and that is the nature of our life. Every corporation budget includes Research and Development. Many men make a living planning change. For most of us, change is a settled way of life. That does not mean we like it, nor that we can stand up for long under the pressure of it.

Several responses to all this technological change are possible. (1) We could cling to the past. Religious people are especially enticed by this option. Christianity believes in a world that cannot be shaken. The gospel declares Jesus Christ as the same yesterday, today, and tomorrow.

[12] See Donald A. Schon, *Technology and Change* (New York: Delacorte Press, 1967).

PLENTY AND TROUBLE

Hence we tend to ignore change, or resist it, or deny it. (2) Or we might seize every new thing as panacea and consider every change as progress, as though it were an escalator taking us ever upward and onward. (3) Or we might give up in frightened resignation. "On earth nations will stand helpless, not knowing which way to turn from the roar and surge of the sea; men will faint with terror at the thought of all that is coming upon the world, for the celestial powers will be shaken." (Luke 21:26 NEB) Such default would betray our basic Christian commitment to live by faith.

(4) For it is exactly when the unchangeable things are known to change that the power of God becomes most available to men. Then they must live by faith or not at all. When the good earth is no longer good, when the human power to destroy exceeds the divine power to create, when birth can be controlled and death postponed and the genetic nature of human nature be deliberately managed, when men can make and change things which heretofore we felt were God's province only, when all the certainties of the past are shaken and the whole stability of the established order is threatened, and everything once secure is infected with change—*then* we do indeed live by faith, or we fall with everything fallen.

A CHRISTIAN CRITIQUE

My personal experience swings elliptically around two texts: "Here we have no continuing city," and "Behold, I make all things new." Those are the situation and the promise. In this technological world where all traditional securities have disappeared, we live by risk. We set out not knowing where we go. Only one thing we are sure of. We will be on the move and change direction many times as we go along. Already we have torn up roots from the accustomed world. We are ordered now to make a second exodus, to leave the place of slavery and humiliation. The whole human race is called to move out of its slavery to fear, drudgery, sickness, and poverty, out from under the whip of necessity, out into the open, promised land. On the way there will be prolonged wilderness where people are tempted to stand around the stationary golden calf of affluent plenty, or to go back to the familiar, if enslaved, customs of bondage. The promised land will not be a garden of Eden. People must battle against strange new customs and many local gods. But here we have no continuing city, and even technology cannot cancel the promise, "Behold, I make all things new."

Provided, of course, that we believe in the future. Really believe. "We cannot reap the fruits of the technological revolution in a demoralized society," said Toynbee on his eightieth birthday.

PLENTY AND TROUBLE

Of course, technology is not the Savior. The Promised Land did not *save* the Hebrews, but it became their home where they had to deal with God, and he with them. Even in the promised land, especially in the promised land, biblical ethics ceases to be an optional luxury. For as Kenneth Boulding says so eloquently,

> In the spaceship Earth, we had better learn to love our enemies, or we will destroy each other. . . . We will have to learn to be poor, or at least parsimonious, in spirit, even in the midst of material affluence. We will have to learn how to mourn, or we will be gobbled up in egotism and pride. We will have to learn to hunger and thirst after righteousness, for there will be no effective sanctions. We will have to learn how to be merciful, for we all have to live at each other's mercy.[13]

Elting Morison intended to say a hopeful word, but he summed up the latent tragedy. "Men have lived with machinery at least as well as, and probably a good deal better than, they have yet learned to live with one another." Unless that is changed, the apocalypse will happen and the snakes take over the garden and worms eat up the apples.

---

[13] *Human Values on the Spaceship Earth*, p. 13.

Suggested Reading

I list here the major works which have been most insightful to me, and which are readily available in this country.

About the Meaning of Technology

Boulding, Kenneth. *The Meaning of the Twentieth Century.* New York: Harper & Row, 1964.
A leading Christian economist reflects soberly about what it all means.

Chase, Stuart. *The Most Probable World.* New York: Harper & Row, 1968.
A leading social critic speculates about the future.

Daedalus: Journal of the American Academy of Arts and Sciences.
Several issues deal with aspects of this problem, written by leading scholars in many disciplines: Winter, 1965, "Science and Culture"; Spring, 1965, "Utopia"; Summer, 1967, "Towards the Year 2000: Work in Progress"; Fall, 1967, "America's Changing Environment"; Spring, 1969, "Ethical Aspects of Experimentation with Human Subjects."

Ellul, Jacques. *The Technological Society.* New York: Alfred A. Knopf, 1964.
The single most penetrating study of the nature of technology; a prophetic statement that disastrous things will continue to happen unless men resist the technical spirit that now dominates them.

Ferkiss, Victor C. *Technological Man: The Myth and the Reality.* New York: George Braziller, 1969.
A sociological and rather hopeful view of human nature.

PLENTY AND TROUBLE

Harrington, Michael. *The Accidental Century.* New York: Macmillan, 1965.

A quest for the spiritual meaning of technology; proposes vast governmental action; redefines work. A rather penetrating analysis.

Kahn, Herman, and Weiner, Anthony J. *The Year 2000.* New York: Macmillan, 1967.

Authors speculate that technology will make for new **mores,** hence for increased alienation of people and decreased influence of religion.

Mesthene, Emmanuel G. *Technological Change: Its Impact on Man and Society.* Cambridge: Harvard University Press, 1970.

The head of Harvard University Program on Technology and Society, and America's chief spokesman in this field, sums up his reflections.

Morison, Elting E. *Men, Machines and Modern Times.* Cambridge: The MIT Press, 1966.

Anecdotal, but very substantial treatment.

Mumford, Lewis. *The Myth of the Machine.* New York: Harcourt, Brace & World, 1966.

History of technological change from 1500. A basic, fundamental book for the understanding of the technological spirit.

Stover, Carl F. (ed.). *The Technological Order.* Detroit: Wayne State University Press, 1963.

Includes statements by Lynn White, Calder, Theobald, Huxley, etc., and the most concise statement by Ellul of his overall critique.

Concerning Ecology

DeBell, Garrett (ed.). *The Environmental Handbook.* New York: Ballantine Books, 1970.

Just that. A popular survey of the problem.

Dubos, René. *So Human an Animal.* New York: Scribners, 1968.

A distinguished microbiologist makes an urbane, elo-

SUGGESTED READING

quent plea to face the ecological issues; the environment we create can enhance or demolish human life.

Ehrlich, Paul R. and Anne H. *Population, Resources, Environment: Issues in Human Ecology*. San Francisco: W. H. Freeman Co., 1970.
A sophisticated treatment of the enormity of this problem, by America's leading expert on population.

Elder, Frederick. *Crisis in Eden*. Nashville: Abingdon Press, 1970.
A serious treatment of the biblical and theological implications of the ecological crisis.

Helfrich, Harold W., Jr. (ed.). *The Environmental Crisis*. New Haven: Yale University Press, 1970.
Contributions by most competent scientists and scholars.

Mitchell, John G. (ed.). *Ecotactics: The Sierra Club Handbook for Environmental Activists*. New York: Pocket Books, 1970.
One of the leading conservationist groups surveys the problem, with strong recommendations about controls and remedies.

Stewart, George R. *Not So Rich as You Think*. Boston: Houghton Mifflin, 1969.
Men have turned the earth into foul dregs and invented linguistic and religious justifications!

Concerning the Machine

Hatt, Harold E. *Cybernetics and the Image of Man: A Study of Freedom and Responsibility in Man and Machine*. Nashville: Abingdon Press, 1968.
A stimulating Chrsitian study of what the subtitle promises.

Concerning Some Biomedical Issues

Augenstein, Leroy. *Come, Let Us Play God*. New York: Harper & Row, 1969.
An anecdotal statement of the ethical issues in genetic engineering.

PLENTY AND TROUBLE

Daedalus. "Ethical Aspects of Experimentation with Human Subjects." Spring, 1969.

The major contribution to clarifying the vast legal, philosophical, and medical questions.

Delgado, Jose M. R. *Physical Control of the Mind: Toward a Psychocivilized Society.* New York: Harper & Row, 1969.

Fascinating and frightening prospects in the control of human thought and behavior.

Ehrlich, Paul R. *The Population Bomb.* New York: Ballantine Books, 1968.

The American expert states the problem vividly.

Rosenfeld, Albert. *The Second Genesis: The Coming Control of Life.* Englewood Cliffs, N. J.: Prentice-Hall, 1969.

A survey of the biological revolution, with strong concern for the social consequences. Popular writing, from original in *Life* magazine.

Vaux, Kenneth (ed.). *Who Shall Live?* Philadelphia: Fortress Press, 1970.

Chapters by Mead, Mesthene, Drinan, Ramsey, Fletcher, Thielicke. Guidelines for decision-making about abortion, transplants, and genetic engineering.

Theological and Ethical Statements

Bonifazi, Conrad. *A Theology of Things.* Philadelphia: Lippincott, 1967.

The most penetrating theological effort to celebrate the unity of man with nature.

Faith-Man-Nature Group. *Christians and the Good Earth.* Alexandria, Va.: Faith-Man-Nature Group, n.d.

A statement by a new group of theologians and scientists committed to Christian research.

Hall, Cameron P. *Human Values and Advancing Technology.* New York: Friendship Press, 1967.

The chairman of the National Council of Churches study project on technology believes the gospel is

SUGGESTED READING

relevant and God is at work amidst the perplexities of this time.

Moule, C. F. D. *Man and Nature in the New Testament: Some Reflections on Biblical Study*. Philadelphia: Fortress Press, 1964.

Wickham, E. R. *Encounter With Modern Society*. New York: Seabury Press, 1964.

A Church of England bishop (leader of industrial mission program) discusses the church's response to technology.

Wilkinson, John (ed.). "Technology and Human Values." *Occasional Paper*, CSDI, 1966. Center for the Study of Democratic Institutions, 1966. Chapters by Wilkinson, Sykes Cabor, Bloy, Roszak, etc.

World Student Christian Federation, "Towards the Future," *Student World* no. 2, Geneva, 1968.

Concerns the ethics of responsibility and the biblical view of the future.

Index

Abundance, 132, 172-78
Achievement, spirit of, 135
Augenstein, Leroy, 108-9, 112, 116

Beckmann, Johann, 8, 28
Beecher, Henry K., 106
Biomedicine, 25, 93-131
Bloy, Myron T., 136
Bonifazi, Conrad, 53
Boulding, Kenneth, 20, 146, 184

Change, 36, 181
Chenu, Maria D., 149
Clarke, Arthur C., 149
Clock, 16
Commoner, Barry, 104
Computers. *See also* Machines
  history of, 68-71
  powers of, 22, 71-76
Cox, Harvey, 50, 66, 87
Cybernetics, 80

Darwin, Charles, 77, 85
Death
  definition of, 114
  ethical issues in, 116
Deats, Paul, Jr., 135, 148
Depersonalization, 140, 168, 171
Desacralization, 48
Donor, injury to, 108
Drucker, Peter, 38, 70

Dubos, René, 46-47, 98, 100, 130, 169

Education, 24
Ehrlich, Paul, 60, 99, 106
Elder, Frederick, 46
Ellul, Jacques, 39-41, 144, 163

Fagley, Richard M., 95
Faramelli, Norman J., 79-84
Fire, 13
Freedom, 134
Future, 19-26, 178-84

Gabor, Denis, 174
Genetic pool, degeneration of, 126
Genetics
  controls, 120
  engineering of, 119-28
  ethical issues in, 124, 128
Glass, 15
Goodman, Paul, 138

Hafez, E. S. E., 123
Hardin, Garrett, 101-3
Harrington, Michael, 139
Hatt, Harold E., 78-79, 83
Hilton, Alice Mary, 143
History, distortion of, 11
Hoffer, Eric, 51
Holden, John, 108-9
Human nature, 66, 171
Huxley, Aldous, 47, 128

INDEX

Informed consent, 107

Knowledge, growth of, 31
Krutch, Joseph Wood, 87

Laboratories, science, 32
Life, simplicity of, 172-78
Luddites, 157

Machines. *See also* Computers
  capacities of, 79-85
  incapacities of, 85-87
  symbiosis of, 88-92
MacLeish, Archibald, 139, 154, 165
Man
  as *homo faber*, 13, 68, 92, 161
  as *homo sapiens*, 68, 162
  as *imago dei*, 131
  self-made, 93-132
  valuing trait in, 161-68
Materialism, new, 59
Mayo, Charles W., 116
Mazlish, Bruce, 17, 77
Merton, Robert K., 39
Mesthene, Emmanuel G., 7, 38, 142, 145, 166-68
Morison, Elting E., 84, 91, 184
Moule, C. F. D., 58, 65
Muller, Hermann J., 127, 130
Mumford, Lewis, 11, 13, 14, 40, 67-68, 170

Nature
  abuse of, 42
  assumptions about, 46
  care for, 58
  Christian attitudes about, 48-65
  cruelty of, 62
  people belong to, 60
  restoration of, 57
  rights of, 54
Novak, Michael, 60

Organ transplantation, 105-19
  alternatives to, 111
  cost of, 118
  guidelines for, 106, 113
  right to sacrifice in, 110

Paradise, Scott, 59, 175
Pollution, 42-46
Population
  controls, 100-105
  explosion, 95-100
Pride, encouragement of, 148
Printing, 15

Quality of life, 150, 177

Railroad, 17
Ramo, Simon, 179
Responsibility, chosen for, 158-61
Rosenfeld, Albert, 94, 122-24
Roszak, Theodore, 164

Santmire, H. Paul, 55, 62
Schon, Donald A., 181
Shinn, Roger, 102
Simpson, George G., 61
Sittler, Joseph, 54, 64, 65
Strain of living, 145-47
Symbiosis, 88-92

Technological assumptions, 163
Technological syllogism, 160
Technology
  ambiguity of, 142, 146, 153, 178
  as inevitable, 170
  avoids moral judgments, 41
  benefits of, 132-38
  characteristics of, 29-37, 166-68
  charges against, 139-52
  critique of, 153-84

Technology—*cont'd*
  defined as, 28, 36, 38-40
  history of, 11-19
  innovations in, 13-19, 35
  mistaken views about, 7-8, 181
  spirit of, 28-37
  spirit of play in, 136
  tool or process, 164
  values in, 161-68
Toffler, Alvin, 36
Travel, 22

Utopia, 178

Values and valuing process, 161-68
Vaux, Kenneth, 112

Weiner, Norbert, 80, 83, 89-90, 146
White, Lynn, Jr., 17, 28, 49, 178
Whitehead, Alfred North, 32
Wickham, E. R., 156
Woodruff, F. A., 113
World Council of Churches, 102, 135, 141, 149, 180